KT-489-704

Craftsman-Style Houses

Fine Homebuilding®
GREAT HOUSES

Craftsman-Style Houses

The Taunton Press

Cover photo: Chuck Miller
Back-cover drawing: Bob LaPoint;
top photo: Chuck Miller; bottom photo: Tim Snyder

Dedicated to Peggy Dutton

for fellow enthusiasts

10 9 8 7 6 5 4 3
Printed in the United States of America

A Fine Homebuilding Book

The Taunton Press, 63 South Main Street, Box 5506,
Newtown, CT 06470-5506
e-mail: tp@taunton.com

Distributed by Publishers Group West

Library of Congress Cataloging-in-Publication Data

Craftsman-style houses.
p. cm.
Fine homebuilding—great houses."
"Collection of articles from the first ten years of Fine homebuilding" – Introd.
Includes index.
ISBN 1-56158-014-7
1. Small houses – United States. 2. Architecture, Domestic – United States 3. Arts and crafts movement – United States–Influence. 4. Architecture, Modern – 20th century – United States. I. Fine Homebuilding.
NA7208.C68 1991 91-15606
728' .373'09730904 – dc20 CIP

Contents

Introduction

The Craftsman-style house was the most popular small house in the United States in the first two decades of the 20th century. And over the past decade, it has also been a favorite of *Fine Homebuilding*'s readers and editors.

Along with Frank Lloyd Wright's Prairie style, the Craftsman style was deeply influenced by the English Arts and Crafts movement. Its principal identifying features are low-pitched, gabled roofs with wide overhanging eaves and exposed roof rafters. But its enduring significance to our built environment surpasses this quick technical summary. In this collection of articles from the first ten years of *Fine Homebuilding* (issues 1-66) you will find renovations and newly constructed houses, decks and island retreats, large houses and small spa rooms, *shoji* walls and stone walls. What unites this collection is what united the outlooks of Charles and Henry Greene, Gustav Stickley, Bernard Maybeck and others — an appreciation for handcrafted natural materials, a love of craftsmanship and a preference for warmth and utility over ornamentation and architectural revivalism.

– *The Editors*

Stickley's Parker house. **Gustav Stickley designed the 'summer bungalow' (drawing at left) for his photographer friend L. C. Parker in Morris Plains, N. J. Its stone entry, which is now obscured by a big fir, leads to a chestnut-paneled foyer. The 20-in. thick first-story walls are fieldstone, collected at the site, and the walls above are wood frame and shingle, with a 10-in. exposure. The low-pitch roof and saddle-bag dormers make for headroom and light in the upper stories, and the whole house successfully radiates what Stickley called 'homelike charm.' The foyer in the 1913 Parker house (above) has floor-to-ceiling chestnut paneling and 4x6 chestnut beams overhead. The window seat on the stair landing offers views outside and a line of sight through the living room, sun porch and yard on the opposite side of the house. The lantern above is a standard catalog item from Stickley's Craftsman workshops.**

Craftsman Houses

Gustav Stickley as architect and interior designer

by John Lively

Carole Harper walks along the edge of the L-shaped upstairs hall, turning on lights inside the rooms and shutting all the doors. Not an odd thing to do if you like hallways as rooms in themselves. The doors are chestnut, frame and panel, with a pair of amber stained-glass lites at the top. Four of them go to bedrooms, the other two to baths. Directly opposite the stairway, where I am leaning against a 6x6 floor-to-ceiling newel post, is a built-in linen closet with double cabinet doors at the top and drawers at the bottom—the whole thing fitted out with hand-hammered copper hinges, latches and pulls. Lit from behind, the stained glass in the doors glows softly, casting a yellow hue on the doors across the hall, on their wide casings, on the cabinetry and on the beams overhead. "Standing here," she muses, "is like living inside a piece of furniture, and you can understand what a Craftsman house is really all about."

This is an apt remark about a house designed by a man known more for his furniture than his architecture. Built in 1913, the Parker House (drawing, facing page) stands at the high-water mark of Gustav Stickley's career as a designer and publisher of Craftsman-style homes. Though never trained as an architect, he began in 1904 to publish designs for houses in his three-year-old *Craftsman* magazine. As a successful maker of Mission furniture and a publicist of Arts and Crafts ideas, Stickley had decided to design houses that would be proper expressions of his democratic philosophy and his Spartan ideas on art and architecture. In short, he wanted to design houses that would write in larger script the messages carried by his furniture.

Origins of the Craftsman Movement—In the history of architecture and design, there is a tension at work between the complex and the plain. After a decade or two of involuted acanthus scrolls and chinoiserie brocades, you get a hard return to stripped-down classical proportions, unadorned line and basic, structural forms. Something like this happened at the turn of the 18th century, when Neo-Classical values won out over the then grown fulsome character of late Georgian design.

But as the 19th century wore on, inventors and engineers expanded the capability of woodworking machines to produce ever more intricate spools, fretwork, carvings and gewgaws. By the 1860s, millworks were spitting out ornate posts, railings, balusters, brackets and the hundred other odd bits of architectural bric-a-brac that embellish Victorian homes. Furniture factories were doing the same thing on a smaller scale (as furniture design almost always mimics architecture) by making chests and tall-case clocks that looked like miniature manor houses and midtown bell towers.

A decade earlier in England, where the Industrial Revolution was already in high gear, enlightened reaction grew up in certain quarters against the aesthetic of machine-made goods, and against what was considered the dehumanization of those who labored in the factories. The two great leaders of this movement were John Ruskin and William Morris, whose prolific writings did a lot to call public attention to the ills of unbridled industrialism, and the uninspired products it cranked out. They called for a return to hand craftsmanship, and for a workshop environment in which the craftsman could design what he made and be happy in the process as well as in the result. For inspiration they turned to the Middle Ages—its guild system of politically important, self-fulfilled artisans, its design motifs and its idealized notions of love, honor and death.

Morris and several others formed a company to make handmade goods available to the common man, but soon learned that paying their craftsmen an honest wage meant charging more than the common man could afford. And so the rich wound up reaping the rewards of Morris' socialist ideals; they were the only ones who were sophisticated enough to share his upstart sensibilities and moneyed enough to afford the pricey things he sold.

Gustav Stickley dedicated the first issue of *The Craftsman* magazine (September 1901) to William Morris. Then with all the moralizing, long-winded zeal of a late Victorian social reformer, Stickley set about to change the American public's taste in residential design, holding up the Arts and Crafts Movement as an example. At the heart of Stickley's own idealism stood the upright American citizen, who valued sincerity over sophistication, utility over ornament. Stickley believed that the real American didn't want more than his fellows, and that every citizen should wish to own things of honest, simple beauty, things that pleased their makers in their manufacture, things that inspired their owners in their possession. Such was the attainable, obligatory goal of a democratic society.

But unlike Morris, who was the son of a well-off London broker, Stickley spent his childhood lugging stones and mixing mortar for his mason father, and he knew from experience that brute toil demeaned more than dignified. And so contrary to Arts and Crafts thinking, Stickley championed the machine. As the Bauhaus people were to do a couple of decades later, he designed furniture and architectural details to take advantage of what machines can do. In his Craftsman workshops woodworking machines abounded—table saws, bandsaws, planers, mortisers and boring machines. He wanted his workers to spend their efforts doing things that machines could not—trimming to fit, assembling parts and finishing wood. He opposed wasting a craftsman's time preparing and dimensioning stock by hand, as Morris' Cotswold craftsmen did. But Stickley never used machines to achieve decorative effects or to produce finished surfaces. These were the special provinces of the hand and eye and the spirit of the worker.

From furniture design to architecture—Not unlike the Shakers, Stickley and other Americans who embraced the Arts and Crafts point of view recoiled from ostentation and superfluous ornament. What the Shakers saw as ungodly and profane, Stickley saw as socially oppressive, tasteless and vain. An early ad for his furniture in the *Chicago Tribune* entitled "Furniture as Educator" claims that his new designs "cast off the shackles of the past" and "help make life truer and better by [their] perfect sincerity." He goes on to say that his furniture has minimal ornamentation, that it emphasizes color and that in form it is "angular, plain and severe." Hardly the kind of claim made by a marketing wizard.

While much of his furniture is indeed angular and severe, his architecture, especially his interior design, is quite simply cozy, homey and warm, without any pretentions to elegance, but equally without the bare-bones severity he claimed for his furniture. Walking through his houses and paging through his magazines evoke the same feelings you got as a kid looking at illustrations in fairy-tale books, seeing how the elves lived in their woodland abodes and imagining yourself having supper there, eating simple fare at a heavy oak trestle table opposite a blazing fire and stone hearth.

As a self-proclaimed architect, Stickley was fairly satisfied with copying the then popular vernacular styles—mostly English country cottages, Adirondack log lodges and Indian bungalows—and never really brought much creative energy to bear on the exterior of his houses. But

Photo: Dick Swift; Drawing: Courtesy of Carole Harper

The living room in Stickley's own house in Syracuse, N. Y., top, includes several elements that he was to use in designing houses for the next 15 years—chestnut beams, posts and paneling, a plastered frieze at the top of the walls and an expansive bay with window seat. The light patches on the wall above the couch and on the post in the foreground mark where original Craftsman sconces were. Fireplaces were the heart of a house in Stickley's view, and the one illustrated above typifies his sentiments. Flanked by built-in benches and bookshelves, the fieldstone fireplace and tile hearth create the kind of intimate, familial air that Stickley thought essential to a proper home life. The partial partitions, open above the wainscot, serve to separate the inglenook from the hallway, but at the same time leave it open to the comings and goings without. Stickley liked rooms that offered a sense of privacy without imparting a feeling of isolation.

Drawing from *More Craftsman Homes*

as a designer of interior spaces, he was in his proper element, and was able to bring what he knew about furniture design into an architectural dimension.

His first real attempt to create an interior from whole cloth was in his own house in Syracuse, N. Y., which was damaged by a fire in 1902. Instead of restoring the house to its original, two-year-old Neo-Colonial condition, Stickley seized upon the chance to redo the inside completely and make it over in the Craftsman image.

Making simplicity work—The dominant features on the main floor of Stickley's Syracuse house are the chestnut beams overhead and the broad expanses of dark, chestnut paneling on the walls. Light plastered surfaces peek out above the paneling and between the beams, as shown in the photo of the living room (top of page). The beams are bold, dark strokes, an effort on Stickley's part to decorate with structural elements, and when combined with the paneling, the whole effect is almost too much. But the large window bay and built-in seat lighten and brighten the room, and leaven the whole-wheat heaviness of all that dark wood.

Touring the downstairs last spring, I was sad to see the rooms stripped of their stained-glass sconces and lanterns. With these to add color and light, and with the original Oriental rugs in place over the stained wide-board oak floor, it must have been a pleasant sight in its day. The Craftsman stamp is everywhere—on the doors, with their three flat panels; on the unmolded, unmitered casings; and on the fireplace with its hand-wrought metal hood.

Looking at the simplicity of the detailing—its self-conscious naiveté, its repetitive use of rectilinear elements and straightforward joinery, you begin to wonder what's so craftsmanly about the Craftsman style. The answer is surface quality—texture, color and finish. To know how important surface quality is to Craftsman-style detailing, you have to see what an awful mess it is when it's poorly done.

Several years ago, I chanced upon a just-completed Greene and Greene style home in southern California. The real-estate broker, who was having a wine-and-cheese open house at the time, was quite beside herself, fluttering here and there and inviting the guests to marvel at the authentic Craftsman features, and hoping that one would be compelled to fork over the immodest sum she was asking. What she and the builders had ignored was surface quality.

The redwood beams and panels, the door frames and window casings were all straight from the thickness planer, which apparently had been set to run the maximum linear footage per minute. Instead of sanding the mill marks out of the wood, the builders had left them there, and the finishers had covered them with several gloss coats of milky, brush-on lacquer. They did this after smearing plastic wood, with what I guessed to be a sponge trowel, in and around the nail holes. Looking into the wood was like looking at decoupaged bulldozer tracks. It was not pretty.

Something of the alchemist lurked behind Stickley's spectacles, for he was a superb concocter of wood finishes, and much of the craftsmanship that went into his houses and furniture was spent preparing, staining and finishing the wood. First, everything was carefully planed, scraped or sanded to remove the knife marks that ripple every board dimensioned by a thickness planer. Next was finish-sanding and staining.

He seldom used pigmented stains, preferring instead to darken oak and chestnut with ammonia fumes, using a very strong 26% solution. For furniture he had fuming booths made of tarred canvas; for houses he recommended putting the ammonia solution out in saucers or shallow pans and closing the room up tightly. The ammonia reacts with the tannin in the wood to produce overnight the dark patina of the ages, without clogging the pores of the wood with oily residues. Because it's the wood itself that changes color, the true character of the figure and grain remain crisply intact.

Stickley had developed a complete catechism of woods and finishes. He decreed what effects certain woods worked, what woods were appropriate to certain rooms and what finishes were correct for those woods. He found, for example, that dilute sulfuric acid applied to Southern cypress produced a soft grey-brown that matched Japanese cypress, which had to be buried to attain that hue. The same solution brushed on birch or beech floors imparted a reddish tone that would not rub off with use.

For interior woodwork that didn't get rough use, he called for a two-coat paste-wax finish after staining, and that was it. For wood floors, he recommended first staining and then applying a single coat of shellac, followed by a coat of wax. I've looked at a lot of unrestored Stickley finishes, and in none of them was there a perceptible film between the eye and the surface of the wood. Just a subtle polish and a friendly patina, inviting to the touch.

Hearths and windows—In designing interior spaces, Stickley paid careful attention to hearths and windows, and was at his best when he arranged these elements in a cooperative relationship with one another. Stickley considered large windows essential to a workable interior. To take better advantage of the light, fresh air and views, he included window seats where they seemed appropriate.

In the Parker house, all of the downstairs rooms have window seats, the most surprising and delightful of them being the one in the entry, on the stairway landing between the foyer and the kitchen (photo, p. 8). You can sit there and be in touch with everything that goes on in the house and in the yard outside, yet it has a reclusive feel to it. You could put your feet up, open up a book and read comfortably for hours, adrift in a world of words, just what Stickley would have wanted.

Private places such as this that keep a connection to the mainstream of activity in a house were important to Stickley. But of greater importance was the area where a family could come together as a whole, and it's not surprising that he considered the living-room fireplace to be the spiritual center of the house. "The big, hospitable fireplace is almost a necessity," he wrote in his introduction to *Craftsman Homes* (1909), "for the hearthstone is always the center of true homelife, and the spirit of a home seems to be lacking when a register or radiator tries ineffectually to take the place of a glowing grate or a crackling, leaping fire of logs." To make the presence of the main fireplace more assertive and congenial, he often included built-in seating around it, or tucked it into a storybook-looking inglenook to be flanked with built-in benches, bookcases and casement windows, as shown in the illustration on the facing page. These built-ins made it easier for the elder members of the family to find comfort in warmth, and, Stickley said, the atmosphere of intimacy they created helped children to enjoy moments that they would remember forever. In the absence of built-in seating, Stickley enjoined his readers to arrange the furniture to "circle the fireplace, and bring by the very structure of the house, the family into intimate, happy relationship."

To give the chimney mass a substantial and warm exterior, he often called for it to be veneered in Grueby hand-made tile, as shown at the top of this page, or brick, and fitted with a metal hood. And he designed many fireplaces in stone, either cut block or fieldstone.

Stickley went so far as to develop a fireplace heater that worked something like a simplified Russian stove, using baffles to route the smoke and hot air around a liner that gave up the heat to an in-house air flow. With bricks at $10 per thousand, he figured that the whole fireplace, including sand, mortar and a mason's labor, would run about $65.

Harvey Ellis, architect and illustrator, worked intensely with Stickley for seven months in 1904, just when *The Craftsman* magazine began to feature designs for houses. Ellis brought a subdued elegance to Stickley's heavier schemes, and he introduced muted colors in the architecture and accessories. At top we see a variation on the inglenook pattern, with recessed window seats and a projecting fireplace of yellow tiles. Above, a living room in a style that prefigures some of Frank Lloyd Wright's Prairie School interiors.

Lighting—The bright gems of Craftsman houses are the numerous sconces, lanterns and fixtures that Stickley designed to be made in his workshops and marketed through his furniture catalog or sold at showrooms in the Craftsman Building on East 39th St. in New York City. Soon after he began making furniture, he felt the need to provide lighting that complemented it, but found commercially available fixtures "absurdly out of place." So he began designing and producing his own, along with furniture hardware of all sorts and fireplace accessories. Most of the hanging lanterns, like the two shown in the photos on the next page, show how much Stickley admired things Japanese. These lanterns consist of a pierced or gridlike copper frame (sometimes polished wrought iron), which is capped with a little pyramidal roof. Stained-glass panels, most often amber, were set into the frames to produce the yellowish luminescence that Stickley liked so much. Lanterns were hung in multiples to light dining tables and living rooms, and sometimes a sconce would sit atop a newel post to guide a midnight wanderer to the kitchen or help a late arrival tiptoe up the stairs.

Built-ins—One of the hallmarks of the Craftsman house is the abundance of built-in furniture. Though Stickley's chief occupation was furniture manufacturer, in the houses he designed, he cut out all but the essential items of movable furniture, and built in everything else so that it would combine with the interior detailing to form a harmonious whole. "I plan the woodwork,"

Harvey Ellis drawings: Courtesy of Beth Cathers

The two light fixtures shown here were made in Stickley's Craftsman workshops. These copper lanterns are typical of the Japanesque lighting he designed and sold as standard accessory items. Once the copper parts were shaped, they were planished (lightly hammered) to produce a textured surface. After assembly, the fixtures were rubbed with pumice powder and aged to achieve their rich brown color.

Stickley wrote, "so that it embraces the built-in fittings, so that every bookcase or corner seat is a part of the development of the woodwork." He added, in all modesty, "In no other way can a house be made beautiful, or the architecture of the interior be complete and homelike."

His plans called for bookcases with paned glass doors, china cabinets with drawers below, like the one shown in the photo on the facing page, built-in seating arrangements of several sorts (mostly window seats and inglenook benches), linen closets, desks and tables built in to alcoves and recessed shelving and storage spaces. These things gave him the chance to transform furniture quite directly into architecture. The built-ins I saw at Craftsman farms, at the Parker house and at his home in Syracuse, N. Y., were executed with the same attention to finish, fit and detail as was the furniture that came from his workshops. The only exception was the drawers, which weren't machine-dovetailed as are the drawers in most of his furniture. Some of the ones I saw were joined up with nailed rabbets. But they were all still in one piece.

Especially important to the Craftsmanly look of these built-ins is the hand-wrought copper or iron hardware made by the metalsmiths in Stickley's workshop. And probably nothing contributes more to the storybook air of his interiors than this hardware—the heart-shape strap hinges for cabinet doors, the V-shape copper drawer pulls with their planished (lightly hammered) escutcheon plates and square-head nails and the wrought-iron andirons and fire tools. They all have a rich brown glow about them that suggests they were forged in some subterranean smithy by fugitive dwarves. Stickley's smiths finished the copper by rubbing it with powdered pumice and letting it age. If they wanted to darken the copper in a hurry, they torched it briefly to get the right tone.

Harvey Ellis—In June, 1903, just before he began to publish house designs in *The Craftsman,* Stickley needed help. He was lucky enough to hire an exceptionally gifted architect and illustrator named Harvey Ellis. Known as something of an enfant terrible, given to drunkenness and eccentric behavior, the profligate Ellis puffed a breath of life and color into Stickley's heavy, somber lines and all-too-often monochromatic interior schemes.

There is a super-substantial quality to much of Stickley's early furniture, as though he built tables to withstand the pounding of gauntlets and sword hilts, as though he built chairs to restrain boarding-school bullies. Ellis lightened Stickley's furniture by making it slimmer and less suited to the knockabout world of mead halls and reform-school refectories. He further gave Stickley's pieces an easy elegance by adding delicate inlaid Art Nouveau designs in copper, silver, wood and brass.

Where Stickley had delineated stout, rectilinear forms, Ellis introduced graceful curves, asymmetrical compositions and pastel shades of purple, yellow and green. His talent meshed perfectly with Stickley's ascetic, overbearing sense of style, bringing to it graceful forms, modest ornament and discreet colors.

Ellis was a collector of Japanese prints, and like Stickley an admirer of Japanese architecture. Using a water-color technique derived from the prints he collected, Ellis did several polychrome illustrations for *The Craftsman,* two of which are shown in the photos on the previous page. Ellis' eye for custom-woven fabrics and rugs, stenciled friezes and curvilinear metal work enriched and refined Stickley's designs. Ellis' death in January 1904 cut short his career with Stickley, and even though they had been

From *Fine Homebuilding* magazine (December 1986) 36:24-29

The china closet at the Parker house is one example of the many pieces of built-in furniture on both floors. Like the other woodwork in the house, it is stained chestnut, with hand-hammered copper pulls and latches. The cabinet's head casing is continuous with the casing for the door to the right. Side by side, the door and the cabinet, along with the plate rail, make a balanced composition.

associated for only seven months, Stickley's work as a mature designer bore the imprint of Ellis' influence, and was improved by it.

Passing to the grey havens—In March, 1915, Gustav Stickley was broke, and he filed for bankruptcy. He just couldn't profitably manage his diverse operations—the furniture manufacturing business, the magazine and the terribly expensive Craftsman Building in Manhattan. What was worse, his furniture sales were flagging from stiff competition (even Sears was selling a cheap line of Mission furniture). His own marketing success had turned against him, he had so popularized the style. On top of all this, he had bought a big hunk of land in Morris Plains, N. J., five years earlier and built a large house there, along with several smaller structures. This venture he called Craftsman Farms, and it was to be a great philanthropic project, the end of which was to take wayward boys and school them in a trade and teach them Craftsmanly values and good citizenship.

The school never happened, and not one boy ever so much as swung a hammer there. It was just another drain on Stickley's energy and capital. He had out-dreamed his bank account, and with the foreclosures that followed he lost everything—the furniture workshops, the farm, the magazine, the building downtown, and all ambition. The curtain fell upon Stickley and never rose again, and nobody much cared.

Stickley retired to his old home in Syracuse and lived there with his daughter's family. They gave him the apartment way up on the third floor. He spent the remaining 26 years of his rather long life puttering around up there, using his cookstove to brew experimental wood finishes (which had to be periodically cleared from the sink traps by a plumber), hobnobbing with his old cronies from the workshop and watching his grandchildren grow up. He was 84 years old when he died in April, 1942.

The legacy—A few years ago, a builder was showing me through a just-completed house in Santa Fe, N. Mex. In the living room there were several Stickley pieces—some side chairs, an end table and a library table. I remarked to the owner that his Stickley furniture was in very good shape. "Who's Stickley?" he asked, unaware that those few pieces would fetch eight or nine grand in SoHo. "Money's not important to me," he said, after I gave him a quick explanation. "But even if the stuff were worth ten times that, I wouldn't sell it. I like it. It's homey. It's me."

The revival of interest in Stickley and his furniture that came about some 15 years ago has done a lot more than create a strong demand for original pieces of Craftsman furniture. What's really important is that the Craftsman revival has reintroduced to us the value of hand craftsmanship and high-quality materials. At a time when technology is pre-eminent and the engineer can find a way to bend almost any material to the designer's whim, it's good to know that there are a few who can still do near-flawless work within the natural limitations and demands of the medium. A few who will mind the details as carefully as the whole. A few who know there's nothing more highly embellished than a simple thing well done. □

This article was written with the help of Mary Ann Smith at Syracuse University, Beth Cathers at Jordan-Volpe Gallery in New York City and Carole Harper. *For more on Stickley and his work, see* Gustav Stickley, The Craftsman *by Mary Ann Smith (Syracuse University Press, 1600 Jamesville Ave., Syracuse, N. Y. 13210) and* Furniture of the American Arts and Crafts Movement *by Donald M. Cathers (New American Library, Inc., 1633 Broadway, New York, N. Y. 10019). Facsimile reprints of Stickley's* Craftsman Homes *and* More Craftsman Homes *are available from Dover Publications, Inc. (31 E. Second St., Mineola, N. Y. 11501).*

Photo: Dick Swift

When I was seven years old, my late grandmother began building a small stone cottage in the forest on our family property in the Maryland countryside. She called this project her "mad house." She worked without plans and alone, save for a single farmhand who helped her gather fieldstones and mix cement.

During the next four summers I watched as the irregular grey walls grew from the forest floor. On those days when my sisters and cousins and I helped to gather stones, we were allowed to dig in the black soil near the trunk of the great red oak that grew several feet from the house. We searched for marbles hidden there by the elves who inhabited our woods. As children, we were too noisy and moved too slowly ever to see the little people. Our grandmother, however, had gained their confidence and was thus able to reveal to us the many elements of their habitat that we could admire, but under no circumstance, disturb.

She showed us the multicolored toadstools that sheltered them from the rain, their soft green moss gardens, and the hollowed-out places in the gnarled trunks of old trees where they sought refuge from dogs and children. We admired their artfully painted lichen chrysanthemums, rendered in orange and luminescent green on the craggy rock outcropping near the forest's edge. We stood on our tiptoes to see the fibrous fungal elves' seat that grew from the exfoliating bark of an old dogwood tree.

My grandmother eventually added a kitchen, bathroom and furnace to her house and moved in. She spent the last 20 years of her life in ever-deepening communion with the natural world, attending the moving seasons, becoming herself part of the pulse of growth and decay.

It was these sentiments about man's relationship to the natural world, first nurtured so well in the imagery of my grandmother's lore and later affirmed and refined by extensive travel in the

From *Fine Homebuilding* magazine (Spring 1988) 45:70-75

Eastern Details, Western Framing

A Japanese-style house in the Maryland woods

by Douglas Hamilton

Shoji wall. **Facing the garden, with views across the lawn and into the woods beyond, is a large expanse of glass shielded by traditional Japanese sliding doors on the inside (above). The cabinet at the far right is a storage box for the screens. They just lift out of their tracks and fit back into the cabinet when a full view outside is wanted. Seen from the veranda (facing page), the rice-paper panels mute and diffuse the light from within.**

Buddhist Himalaya, that spawned my interest in architectural elements that form a symbiosis with their environment.

In 1972, during my first Asian trip, I had the good fortune to meet Tsognie Wangmo, a woman from the Himalayan Kingdom of Sikkim. After a lengthy long-distance courtship, Tsognie consented to become my wife. We were married in 1974, and Tsognie came to live with me in Maryland. She brought with her, by way of her Mahayana Buddhist upbringing, a profound belief in the transitory nature of life, of man's place in the ephemeral realm of *samsara*. Although our backgrounds were completely different, we found that our values were quite similar. We had both come to believe, via different paths, that a person's physical environment inexorably molds his values and facilitates or hinders his acceptance of the cycle of life.

Early in our marriage we chanced upon a book of photographs of Japanese gardens and houses. These images evoked a strong sense of déjà vu in both of us, and we began to collect more books on the subject. We recognized that the environmental and cultural factors that had led us to appreciate Japanese architecture were the same elements that had influenced the development of Japanese aesthetics centuries earlier.

Our Himalayan travels had given us two other archetypes. Sikkimese village architecture follows forms that are similar to the more primitive structures found in rural Japan. The use of bamboo, unadorned wood, wide roof overhangs, stucco walls and verandas are common to both.

The other example, surprisingly enough, came from the British, who established a strong presence in the Himalayas beginning in the mid-1800s. They developed a network of "dak" bungalows for the use of traveling government officers in the hills of West Bengal and Sikkim. Typically, these bungalows were single-story structures positioned to take maximum advantage of the view. Like the vernacular Sikkimese houses, they featured covered verandas, well suited to the four-month Himalayan monsoon. Many of these bungalows are maintained today, and we learned to appreciate their utilitarian simplicity while trekking through the region.

Finding a site—By the late 1970s, Tsognie and I were in a position to pursue our dream of developing a permanent home. With my cousin Bruce Hamilton, who also had an interest in rural land, we toured the rolling horse country of northern Baltimore County, seeking a building site with a reasonably mature deciduous forest, a running stream, some tillable acreage and a good hill for sledding.

In late 1979 we bought a 35-acre tract that had been carved out of a large farm. The property rose high on a southern slope, looking across the Gunpowder River Valley to the hills beyond. The site that Tsognie and I chose lay on a knoll in the center of a beech grove. We could stand on the hard silver trunk of a fallen locust tree and easily envision the view from every position within a future dwelling.

In the ensuing two years we carefully removed trees and underbrush to open the site and give a better view of the surrounding area. It remained a top priority throughout the development process to protect the beeches, whose shallow root systems are particularly vulnerable to soil compaction. No piece of earth-moving equipment ever worked near these trees without my being present to control where it was able to go. This added considerably to the cost of site work, but was money well spent.

Finding architects and builders—In early 1981, with only a vague sense of what constituted a "Japanese" house, we began to search for an architect who could combine our sensi-

Kitchen/family-room wing. **The Douglas fir beams and ceiling decking in the kitchen (left) were hand planed, and they work well with the cabinets, which are stock items. Viewed from the porch off the master bedroom, the kitchen/family-room wing (above) sits atop the guest room. The cedar planking for the outside decks was also hand planed and left unfinished. Floor plans are shown in the drawing on the facing page.**

bilities with the technical and aesthetic considerations necessary to design a successful dwelling. We were fortunate to connect with New York based architect Barbara Sandrisser. It didn't take long to decide that she and her partner Peter Paul, both of whom had traveled and studied in Japan, would design our house. Sandrisser's enthusiasm over the prospect of desiging a Japanese house was infectious. She talked of intuition in the design process, which at once struck a familiar note, as I remembered the construction of my grandmother's house. I felt that we had found a kindred spirit.

At the same time, we were also lucky to be contacted by two local craftsmen who had been trained in Japanese carpentry. Carl Swensson, who had recently moved back to Baltimore from the West Coast, had served an apprenticeship with a Japanese teahouse builder in California. Although Swensson, like us, had never visited Japan, he had developed a high level of skill in Japanese woodworking techniques. Paul Wexler had lived and traveled extensively in Japan, and had for a time worked with a revered Japanese temple builder.

Thinking through the design—The design process began in the spring of 1982. We wanted a house that was Japanese from the inside out. We definitely wanted to avoid building a conventional house and adding gratuitous details to give it an Oriental look. But at the same time it was impractical to try to construct a typical Japanese house. We wanted better protection from the elements than the traditional Japanese house, with its paper walls, could provide, but did not want to be isolated from the change of seasons and the marvelous mood changes of Maryland weather. The house was to serve as a pavilion from which we could view the surrounding forest. We wanted to avoid a "front" and "back" to the house, and to devote as little space as possible to the utility area. The garage, to be constructed at a later time, would be located 75 yards up the hill, near the approach to the site.

Materials were to be natural where possible. Wood was to be left to weather; stone and other materials were to relate to the texture and colors of the site, particularly the smooth grey bark of the beech trees. We liked the idea of a house with a center, a courtyard onto which many of the rooms would face. Since we wanted several gardens next to the house, we had to consider how these areas would be viewed from within the house, and how light, weather and time would affect them. We wanted no air conditioning, but rather flow-through ventilation, and as many passive-solar features as the site and other considerations would allow.

We had asked Sandrisser and Paul to design a two-story house, and in January of 1983, they provided us with plans that were sufficiently detailed to allow us to secure some pricing estimates. During the design phase of the project they had worked closely with Swensson to determine the elements of traditional Japanese joinery that would be incorporated into the house. As his work was quite expensive and time-consuming, we allocated only the most important areas to his skillful hands. Swensson worked closely with Sandrisser and me in designing these elements, the result of which were details that would stand on the beauty of the wood and Swensson's skill rather than on gratuitous complexity or decoration.

Swensson would build the south wall of the living room, complete with *shoji* (photo previous page), and the structural beams for the kitchen wing of the house (photo above left), which was the only approximation of traditional post-and-beam work that was incorporated into the design. He would be responsible also for the main entry, closet doors and several small windows. Wexler would build the important veranda (photo p. 14) and the porch and wooden steps that lead to the informal entrance (top photo, p. 18).

It was at this point, while securing pricing, that we faced our only major setback. We found that we could not afford to build a house of 3,000 sq. ft. and still maintain the level of quality we wanted. So we asked Sandrisser and Paul to redesign the house as a single-story structure. This reduced the square footage, but did not significantly change the footprint of the house. It did call for substantial structural changes, and it altered the line of the roof.

One of the positive developments to come out of the bidding process was Paul's recommendation to use a construction manager rather than a full-fledged contractor. Local builder Jack Day agreed to serve as construction manager for the project, which meant that he would take responsibility for 90% of what a contractor would do, other than paying the various subcontractors. That would leave me free to coordinate the special areas of concern—the Japanese carpentry, the stonework and the ceramic-tile roof. Day would work for a fixed fee, and provide at my cost an on-site construction superintendent to coordinate the work on site.

Prior to the commencement of construction, in accordance with Sikkimese tradition, Tsognie's *lama*, or Buddhist priest, visited Maryland to bless the house site. This is done to pacify any malevolent forces that may be in residence and to ensure harmony for the future household. Dodrup-chen Pinpoche found the site to be most auspicious when he blessed it in early spring, and he presented a ceremonial *kata*, or silk scarf, to the land.

We broke ground in early summer. The one-story design actually included space on three levels (drawing, right). As is customary in Sikkimese houses, Tsognie's "shrine room" was designed to occupy the highest point in the house, a small second-story loft. Today, Tsognie maintains her Buddhist shrine in this area, a place set apart from the rest of the house to practice meditation. A special sitting area lets her view the large straight trunks of the poplar grove that grows near this side of the house. This simple view constitutes her own Zen-type garden.

The challenge of working with a steeply sloping site gave us the opportunity to develop space under part of the main floor that is substantially out of the ground. Sandrisser's very careful positioning of the house in a southeasterly direction lets light saturate this room for much of the day, especially in the winter.

Because the house is roughly in the shape of a U, with the courtyard in the center, every major room has at least two walls open to the outside. The kitchen/family room and the master bedroom have three. In the winter, when the sun is lower in the sky, sunlight bathes every room with light for most of the day. In January and February, our coldest months, the sun alone is sufficient to raise the temperature within the house to comfortable levels. In the summer, when the sun is higher in the sky, the combination of large deciduous trees near the house and the 3-ft. to 9-ft. roof overhangs prevent the sunlight from directly entering the house.

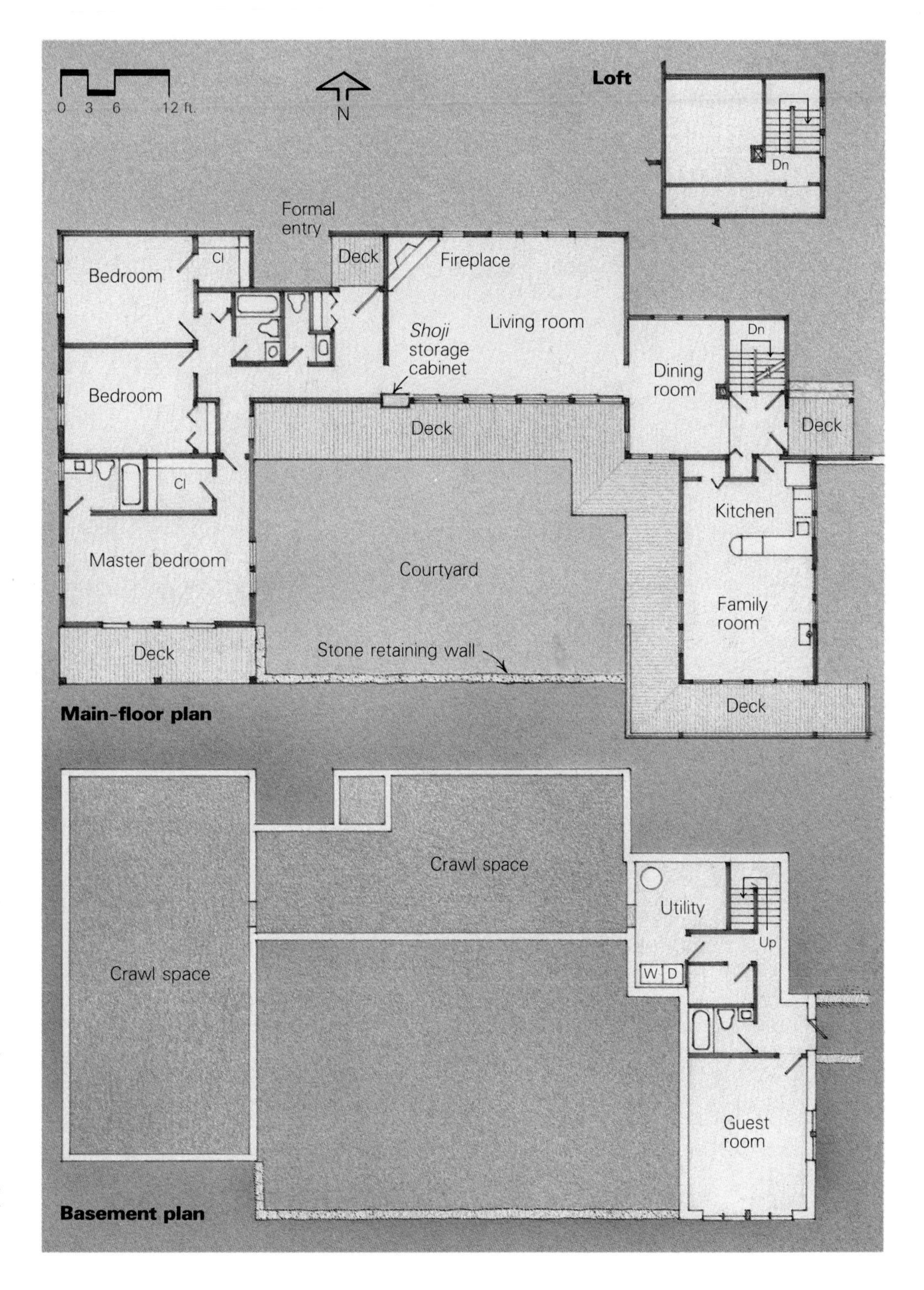

Exterior finishes—We gave very careful consideration to the exterior finishes for the house. We wanted the house to be rooted to the site, and for us this meant stone. Specifically, it meant the same stone that my grandmother had used, although the method of laying the stone would be different. Sandrisser and Paul incorporated a stone base around the entire perimeter of the house. Day was able to locate a superb stonemason in a nearby county. Russ Tagg, a third-generation mason, referred to his skill as "the family curse."

The stone we wished to use was a fieldstone granite, grey-black in color and very irregularly shaped. As it is not sedimentary rock, it does not break along even planes, and must thus be fitted together as it is found. In our case, this was made even more difficult because we did not want to see any mortar. Tagg visited several farms near my family home to examine old stone walls. Based on his confidence that he could work with the stone, we purchased two walls and had them hauled to our property, about 25 miles away. We used a total of 120 tons in the base of the house and the nearby retaining walls. Our stone cost about $18 a ton, delivered, far less than the more typical $70 to $100 for stone purchased from local quarries.

One advantage to not breaking the stone was that the surface patina of the rocks was not disturbed. This had the effect of giving the wall a settled look as soon as Tagg fitted the complicated jigsaw puzzle together.

We selected Port Orford cedar for all the exterior fascia boards, verandas, soffits and trim. As this wood weathers, it turns a handsome silvery grey that will relate well to the beech trees, the stone and the ceramic-tile roof.

Wexler hand planed each board in constructing the verandas, which form the transition spaced between the interior space and the gardens. Most of the verandas are covered by wide roof overhangs; these let us sit outside and enjoy the rain, which we both consider to be perfectly fine weather.

Interior Japanese carpentry—Swensson began working on the *shoji* wall and kitchen beams before groundbreaking. After Sandrisser and I refined some of his ideas, he set to work pur-

Drawing: Gary Williamson

***Entries.* Designed without any 'front' or 'back' in the usual sense, the house has two entries—one formal, the other less so. The formal entry with foyer (below left) has a coat closet. The entry door is custom made from cherry and mansonia, and the casings for doors and the small window are Port Orford cedar. The informal entry (above), with a plain cedar door and small deck, is to the side of the house, and gives immediate access to the kitchen and stairway to the guest room below.**

chasing his materials. This proved to be fairly difficult, as he and I insisted that all the material used in his work should be of the highest possible quality. Finding the structural beams and purlins for the kitchen wing turned out to be the most troublesome. We needed three 4x12 Douglas fir beams, each 24 ft. long. Swensson had to drive to a wholesaler in Philadelphia to locate this wood. It had then to be shipped to Baltimore, where it was planed, transported from the planer to Swensson's workshop for his joinery, and then moved again to the job site. He carefully wrapped each beam and purlin in white paper, and then in plastic. It was not until eight months after installation that we were able to remove the protective wrapping and fully admire the handiwork.

Swensson and I chose each board, and often argued the relative merits of a particular piece. We used a combination of cherry, Port Orford cedar, Douglas fir, and mansonia for various elements. Where possible, Swensson used only hand tools. All interior woodwork was left unfinished, although the hand-planed surfaces have a patina that makes them look lightly varnished. In most locations, Swensson used only quartersawn wood, which adds greatly to the strength and stability of the finished product.

Very early in the job it became apparent that Swensson had underestimated the time it would take to complete his part of the work. This meant finding temporary solutions to closing in the house. When we moved into the house in late winter, 1984, we had temporary metal doors and plywood panels nailed over window openings. It would be almost two years before all of Swensson's work would be completed.

A tile roof—One of the strongest features of the house, and potentially a major threat to the budget, was the roof. We investigated domestic ceramic tile and slate, and found them to be both expensive and aesthetically unsatisfactory. Sandrisser pulled a bit of a coup by finding an importer of traditional Japanese tile (International Tile and Supply Corp., 1288 S. La Brea Ave., Los Angeles, Calif. 90019) that was able to take our plans and provide a favorable price. When local estimates to do the installation proved to be prohibitive, the importer recommended a Japanese installer who lived in California. For a fraction of what local roofers had quoted, Ted Chida brought his crew, which included himself and two helpers, to Maryland to install the roof. The roof weighs over 50,000 lb., a considerable load that had to be structurally accommodated in the design.

Entries and gardens—Unlike Westerners, the Japanese do not have "front" or "back" doors. Just as they avoid symmetry in design, so do they avoid an approach to a house that is too direct, too revealing. Our site lent itself to the more oblique approach favored by the Japanese. We have formal and informal entrances, which are distinguished more by the intensity of the approach and the richness of the material than by more obvious features. To reach our formal entrance, one follows a gravel and stone path that traverses the hill directly above the house. The formal entry door is of cherry and mansonia; the informal entrance door is of fir.

The design and construction of the various paths and gardens are ongoing projects. On two occasions we have hired large loaders to help us move garden rocks from the ridge of a neighboring farm. Each has been carefully selected for its patina, lichen or interesting shape. Just as Tagg worked his jigsaw puzzle, we have tried to form the most interesting arrangements.

Tsognie and I and Daniel Sickles, a local gardener, have collaborated on all aspects of the garden plans, in consultation with Sandrisser on some of the broader issues. As with the house, the selection of harmonious material is a continuing challenge. We decided that a "turkey finisher" crushed gravel was most appropriate for the paths. This gravel, which is fed to mature turkeys to help them digest their food, is supplied to local poultry growers. We were interested to discover, on a recent trip to the Zen gardens of Daitoku-ji in Kyoto, that it is identical to the gravel used in Japan.

Even after three years our house is not complete. It is still in the process of "weathering in" to the site. The gardens, while well under way, are immature and only partially planted. A variety of smaller woodworking projects await our attention. But the way in which we live in the house has achieved a certain completeness.

As the house weathers and changes, so too, do we. Our children watch the maturing tadpoles in the courtyard pond and ponder the mysteries of life. They know the restless smell of honeysuckle and rotted leaf mold carried by the vanguard wind of an approaching thunderstorm.

And usually, it is not the house that Tsognie and I see in our mind's eye. It is suspended sunlight, streaming through partially open *shoji*, forming a chiaroscuro of convergent patterns on the pale white walls of our great room. And it is the trees and sky beyond. □

Living room. **Certain areas of the house feature costly craftsmanship, but others are quite ordinary in terms of their construction and finish. The living room, with its angled fireplace and untrimmed casement windows, is a good example of using conventional finishes effectively.**

Deck details. **The lapped cedar handrails turn upward slightly at their ends, and the midrails are tenoned into the post. The cedar decking is attached with stainless-steel screws from below so that no fasteners show on the surface.**

Greene and Greene Revival

Imagination and careful detailing make a barn a bungalow

by John Lively

Rodger Whipple had been a builder for ten years when he and his wife Susan came to Santa Barbara, Calif., and bought a defunct dairy on a seven-acre hillside lot. Like most builders who finally get around to making their own shelter, Whipple knew what he wanted. He didn't have anything specific worked out on paper, but he did have the germ of an idea. It was California Craftsman, in the mode of Charles and Henry Greene.

Whipple had spent almost seven years repairing and reconstructing Pasadena bungalows (including some work on the Gamble house) and adapting Greene and Greene designs in new construction projects. During this time he developed a feel for how the Greene brothers did things, and became conversant in the language of their architecture. So when, at last, he set out to build his own house, it was natural for him to use their terms to make his own statement about how to put a roof over one's head.

Looking at Whipple's house now, with its coherent design and thoughtful detailing, you wouldn't suspect that it started out as a barn. But it did. He originally planned to build a barn first and use it as a shop for his custom carpentry business. Later he would build his house, higher up on the hill. And while all this construction was going on, he and his family would take up temporary residence—two years, it turned out—in a mobile home.

The barn got as far as its slab foundation when the Santa Barbara zoning officials told Whipple that local restrictions forbade his putting up a new building for commercial purposes at that location. Further, he learned that his chances of getting a variance were slim. Applying for one would be costly, time-consuming and risky. One objection from a neighbor and the whole thing was dead. Whipple decided not to try, and so was stuck with a concrete slab for a barn he couldn't build, and a business he couldn't operate.

Things seemed pretty hopeless until he found out that the existing barn on the property was classified as "legal non-conforming." That meant it was lawful for him to use it for his woodworking shop. Though the old barn would require some fixing up and considerable structural repair, it came as a big relief for him to know that he could make a living at his new location. But he still had a vacant, barn-size concrete slab on his hands.

There was only one thing to do—build the house where the barn was supposed to go. He had liked the look of the proposed barn, with its modular plan and its lapping roof planes. After closely studying the drawings, Whipple found that he could create pleasing, well-related living spaces within the planned modular layout. In fact, he never had another set of plans drawn, but penciled in his changes on the blueprints of the barn. And one of the remarkable things about the house is that he was able to arrive at a workable two-level living arrangement, while preserving the basic footprint of the original plan.

The plan for the barn—A typical California cord-beam barn gets its name from the two continuous beams that run the length of the building. This kind of barn is a straightforward, symmetrical structure, consisting of a long central bay flanked by wings, which are usually divided up into stalls. Above the central bay is a loft that is covered by a gable roof with wide eaves. The side wings have shed roofs that tuck in under the overhangs of the

Built from plans intended for a barn, this Craftsman-style house combines wood, stucco and stone on a hillside site in Santa Barbara, Calif. A portion of each side of the gable roof was raised to create dormers on the second story (photo above), turning a loft into living space. The diagonal braces seen at the end of the house were added to meet code requirements for shear and racking resistance. At the opposite end of the house (photo facing page) these braces are incorporated into a trellis.

From *Fine Homebuilding* magazine (August 1984) 22:54-60

Mixing visible timbers with hidden 2x framing **was typical practice for the Greene brothers, and a method Whipple found economical and structurally sound (left). He painted the framing members and underside of the roof decking before assembly. Only the garage (above), which will become the living room, was framed up mostly post and beam. The lower beam atop the first corbel is one of two cord beams that run the length of the house and tie the structure together laterally. Its sections are linked by double-keyed scarf joints. Corbels were shaped as shown below.**

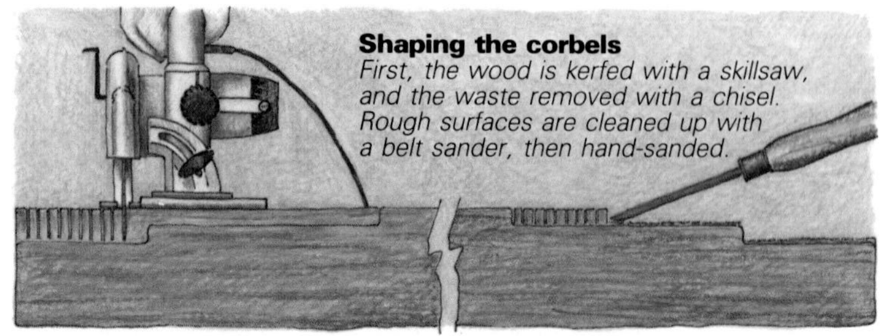

main roof and die into the walls of the loft. The pitch for the gable roof is the same as the pitch for the shed roofs, and the visual effect is of a single gable roof whose plane is broken horizontally and stepped down like a terrace.

Whipple's plans for the barn called for a center bay 16 ft. wide and 72 ft. long, with 8-ft. wide wings running the length of the bay on either side (drawing, facing page, top). The structure's vertical framing would be 6x6 posts placed every 12 ft. along the perimeter of the slab. Another two rows of posts 12 ft. o. c. would run along a line 8 ft. in from the long edges of the foundation, and define the center bay. These would support the cord beams and the loft and gable roof. The cord beams are 6x8 continuous members (the sections are linked end to end with double-keyed scarf joints). They tie the structure together laterally and carry the joists for the loft and the wall framing above.

The loft walls would also be framed with posts. They would support the plates and the 4x6 rafters (32 in. o. c.) for the gable roof that covers the central loft area. The upper ends of the shed-roof rafters would tie into the framing of the loft walls.

From barn to house—The layout of posts and beams created a grid with modules 12 ft. by 8 ft. (see the plan drawings on the bottom of the facing page). When he adapted the plan for his house, Whipple was able to locate load-bearing partition walls along the original grid lines and end up with a plan in which only two posts stand in open floor space: one in the garage, the other in the family room. In the barn's central bay and south wing he located the master bedroom, the family room and the dining room. The garage occupies the east end of the central bay and part of the north wing, while an office fits in next to the garage in the south wing.

The kitchen and foyer wouldn't fit on the original grid, so Whipple broke out of the modular arrangement and added an area 6 ft. by 24 ft. to the north wing to make the needed space. To the west of the kitchen the master bathroom occupies a single module, as does the pantry on the kitchen's east side.

On the house's second story, in the space that in the barn would have been taken up by the loft, he fit a bathroom, two bedrooms and a squarish hallway that accommodates the stairwell, the landing and a window seat. To bring light and air into the hallway and the west bedroom, Whipple raised a 28-ft. long section of the gable roof on the south side, creating a low shed dormer with four windows. He did the same thing for the east bedroom by elevating a 23-ft. long section of the north side of the roof.

Framing—Following the Greene brothers' practice of integrating hidden balloon framing with visible big-beam framing, Whipple used 2x stud-wall and roof framing except for the members that show. For these he used big sticks—Douglas fir 6x8s and 6x6s for the posts, beams and corbels; redwood 4x6s for the rafters and rafter tails. The garage (a space that will in the future become the living room) is framed up almost entirely post and beam, but the rest of the house is platform-framed with posts and beams placed at structurally critical locations.

The two 72-ft. long 6x8 cord beams are visible throughout their lengths, even where they run atop partition walls. Where the beams intersect interior walls and end walls, they sit on 6x6 corbels, which in several places are so long that they are beams themselves. The corbels are spiked onto 6x6 posts that were framed right into the 2x stud walls. The cord beam was notched to receive the walls' top plates and to accept the cripple studs that are nailed to the sides of the posts. Because the posts are wider by 2 in. than the wall framing, they stand proud by ½ in. on each side of the finished wall, and so become decorative as well as structural elements.

Greene and Greene notwithstanding, local building inspectors looked with suspicion on Whipple's hybrid framing. They said it didn't have enough shear value to pass code requirements for racking resistance, and that he would have to beef up all the timber-frame joints with steel gussets and fabricated connectors. Whipple wasn't at all pleased at the prospect of having to bore the timbers full of holes and hide his carefully articulated joinery behind ugly metal connecting plates. It just wasn't acceptable. And so he found himself at loggerheads with city officials again.

After talking with an architect, Whipple proposed buttressing the ends of the building with diagonal 6x6 braces. These, his architect calculated, would provide as much resistance to racking forces as the required hardware, possible more. The braces (photos pp. 20-21) pick up any lateral loading from the cord beams and transfer it to the foundation slab. On the west end of the house the braces look to be part of the trellis and its support framing. But on the east they are noticeable indeed, and here they work visually to anchor the house to the site.

With the exception of the garage, the roof was framed with 2x6s 16 in. o. c. (to take the load of a Japanese tile roof to be installed at some future date). The 4x6 redwood rafter tails, which punctuate the eavelines of all the house's six roof planes, are sistered onto every other rafter, and extend only about 3 ft. past the top plate toward the interior.

Detailing the timbers—One of the hallmarks of Greene and Greene design is the decorative treatment of structural members. Posts, beams, corbels and headers were given careful attention. To soften the edges, Whipple used a ½-in. piloted rounding-over bit in his router. Where timbers come together, the joints are made emphatic by rounding the mating edges with a ⅜-in. bit. The effect of this is especially evident in the keyed scarf joints and where posts and corbels meet.

The corbels, beam ends and gable-end purlins were embellished with cloud-lift motifs. The drawing on the facing page shows how Whipple was able to make fairly quick work of these by first laying out the design on the timber and then scoring the wood across the grain with his skillsaw set to a depth that was just shy of the bottom of the layout line. He chiseled away the remaining waste and then smoothed the resulting rough surface with a belt sander. He rounded over all of the 90° edges, which left everything ready to finish, except for the inside corners formed by the steps in the cloud lifts. Here the router bit left radii that said router bit a little too loudly, so Whipple took rasps and files to these junctures and turned the radii into little crevices.

All of the exterior woodwork was stained before assembly. Whipple used Olympic stains, and then coated over them with Watco clear exterior oil finish.

Curving the roof planes—Greene and Greene borrowed a number of motifs from Oriental architecture. One of these was the penchant for curving roof planes upward at their gable ends. Whipple wanted to treat his

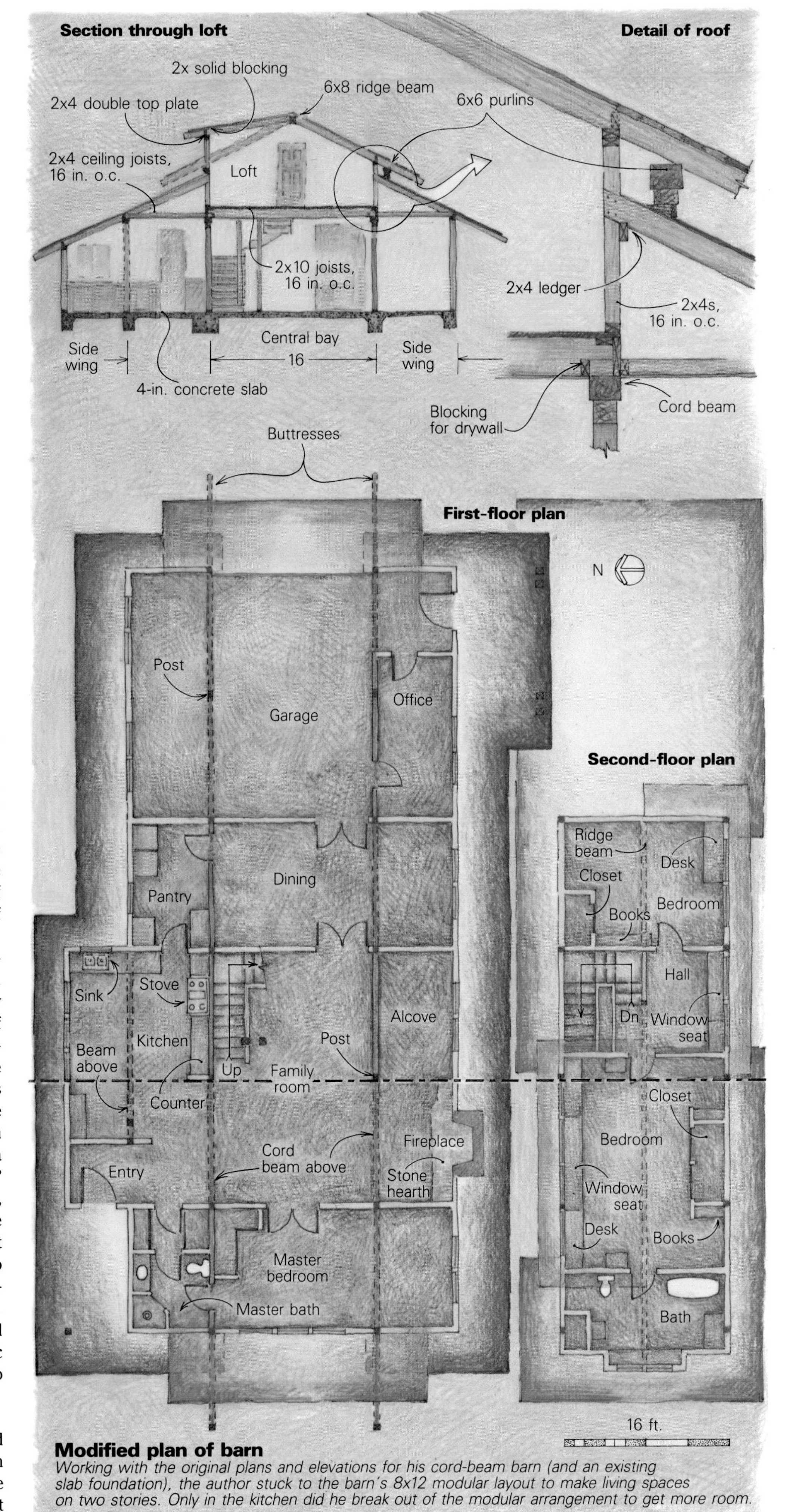

Modified plan of barn

Working with the original plans and elevations for his cord-beam barn (and an existing slab foundation), the author stuck to the barn's 8x12 modular layout to make living spaces on two stories. Only in the kitchen did he break out of the modular arrangement to get more room.

Illustrations: Christopher Clapp

roofs in a similar way, but recognized that too much uplift would war with the house's dominant horizontal lines. So to get the subtlest of upturns at the gable ends of the main roof planes, he applied a shim plate (painted a contrasting color) on top of each of the outermost rafters, making them ⅝ in. higher than the others. This meant that when the 2x6 hemlock decking was nailed down it would curve almost imperceptibly upward between the last two rafters. Instead of securing the overhanging deck with a barge rafter whose visual weight would offset the restrained uplift, he nailed a tapered verge board to the end-grain edges of the hemlock planks and ran a 4⅜-in. by 1⅜-in. barge board flat along the underside of the decking.

The dormer roofs are curved upward by making the seat cuts progressively ¼ in. shallower on the last three rafters at each end. This produced an eaveline that looks to be a gentle sweeping curve from end to end.

Whipple knew it would be a long time before he got around to installing the several thousand Japanese tiles he'd ordered for the roof. So to make do during the several intervening years, he used red 90-lb. roll roofing in the field and bordered it with black.

Interior detailing—Greene and Greene loved wood, and they found ways of using it that took into account its natural appearance and its hygroscopic nature. And while their designs are fairly sophisticated in the aesthetic sense, they are simple and forthright in the practical, workerly sense. The Greenes manipulated lines and planes, light and shadow to produce pleasing effects. They got these effects by segmenting and layering standard rectangular-section lengths of trim. They seldom used ornate, curved-section moldings, and almost never called for mitered corners in their trim, or for flush-fit connections. Instead of concealing joints, the Greenes accentuated them, making them part of the design.

All of Whipple's interior woodwork (except for the kitchen cabinets, which are mahogany) was milled in his own shop from conheart (construction-heart) redwood, which is significantly less expensive than clear-heart redwood, but flawed here and there by knots and checks. He bought about 6,000 bd. ft. of it, and because it was green and roughsawn, he stickered it under cover until it was time to use it.

Whipple's shop is well equipped for the kind of joinery and millwork his project required. It has a small (12-in.) thickness planer, an 8-in. jointer, a nice little 10-in. Wadkin table saw, a radial-arm saw and a drill press. The only really specialized machine in the place is a double-spindle horizontal boring machine, which he used to dowel-join door frames.

Whipple finished the interior woodwork with three coats of Varathane varnish, sanding lightly with 220-grit paper between coats. Nail holes, not all of which he's gotten to yet, will be filled with shellac stick—the kind that's melted into the holes with a hot spatula (called a burn-in knife).

Built-up wainscot—The most expansive wood surfaces in the house are the wainscots in the family room, kitchen and dining room. In the family room and kitchen, the wainscot (facing page, top) has a panel-and-batten look rather than the more conventional frame-and-panel arrangement. The design called for two rows of 8-in. wide, ¾-in. thick panels—a row of long ones at the bottom and a row of short ones (only 4½ in. high) at the top. These were nailed directly to horizontal bands of blocking between the studs with #6 finishing nails, and spaced a uniform 1½ in. apart on all sides.

Next came the three horizontal bands—a 3½-in. wide head band, a 2½-in. wide band that runs 4½ in. below the head band, and a 5½-in. wide base strip. These were all milled to a thickness of ¾ in., and they lie directly over the panels. Instead of mitering the horizontal bands where they meet at outside corners, Whipple finger-joined them in typical Greene and Greene fashion. After the bands were nailed into place, the ⅜-in. thick vertical battens were centered over the gaps and nailed into the blocking. Before any pieces were nailed up, their edges were eased with a ⅛-in. rounding-over bit.

Thus Whipple created a multi-layered surface with a reveal wherever two pieces of wood come together. The bands, where they die into the ⅞-in. thick door and window casings, stand recessed ⅛ in. The battens are ¼ in. thinner than the bands, and the plane of the panels is ⅛ in. beyond the battens. Looking at the wainscot, with its soft lines and relieved surfaces, you'd think that there was more going on here than simple layering and butt joinery. The nice thing is that there isn't.

The family-room fireplace was laid up with cut sandstone blocks quarried on the site. The cord beam sits atop a corbel where it rests on a post (left, outside frame of photo) and on the right where it intersects the end wall.

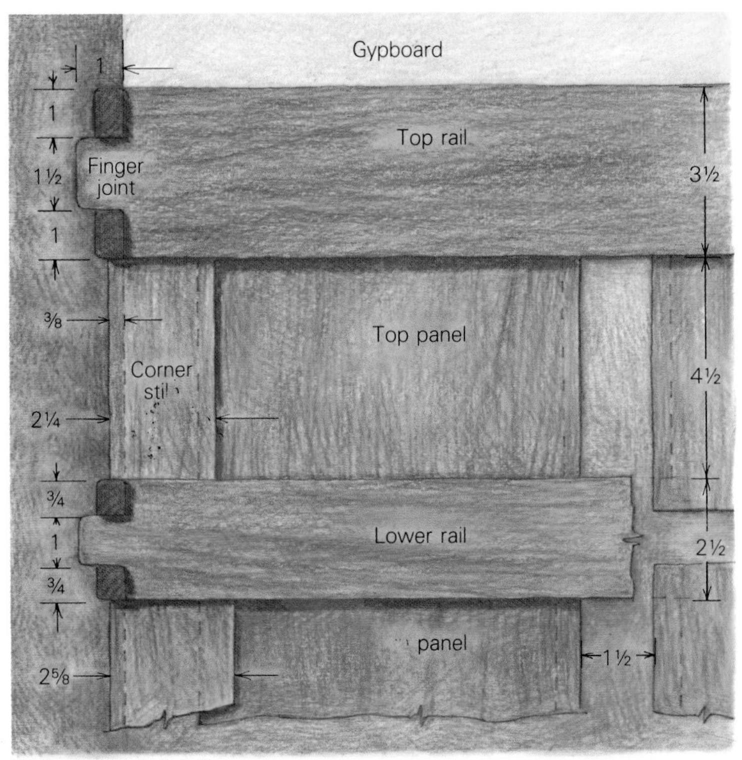

Wainscot and detail

Plate rail and detail

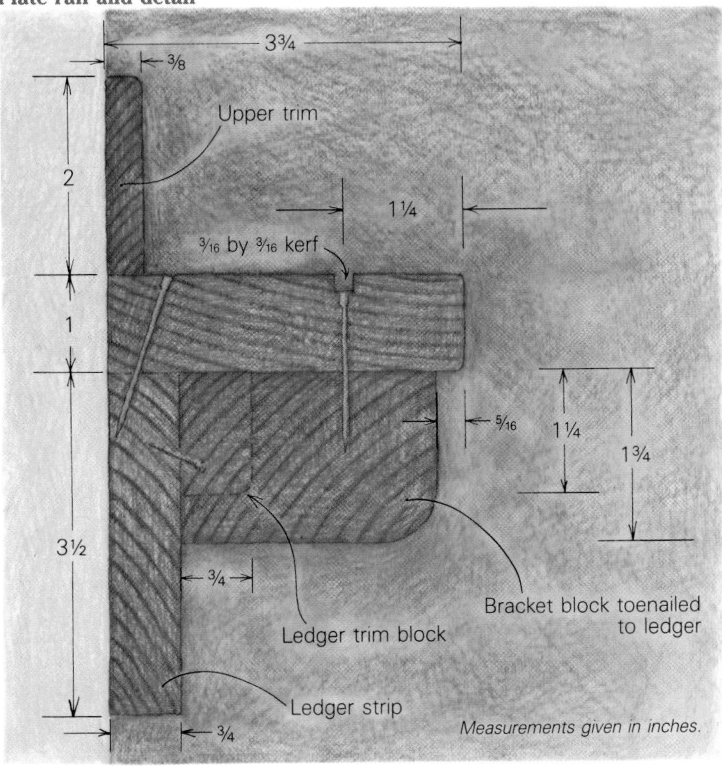

Wainscot and plate rail—In the dining room the wainscot is treated a little differently. It's topped with a plate rail, and lacks the intermediate band and the batten-to-panel proportions of the family-room wainscot. The plate rail (drawing and photo above) consists of five basic parts—a 3½-in. wide ledger strip, a secondary trim band that is interrupted every 14 in. by a bracket block that suggests a beam end, the grooved plate rail itself and a thin band of trim on top. Like the other pieces of composite trim, all the parts of the plate rail were butt-joined and rounded over on their edges before installation.

Doors and casings—Whipple made all the redwood doors for the house in his shop. While these doors look like complex compositions, they were actually fairly easy to make. The outer frame members were planed 1⅜ in. thick, and the inner frame members, which were treated with cloud-lift patterns, were planed 1¼ in. thick. Next, all of the inner edges of the rails and stiles were grooved 1 in. deep to accept the ⅝-in. thick panels, and to house the inch-long tenons that were cut onto the rails. Then the cloud-lift patterns were created by ripping ½-in. wide strips from the boards in the right places and stopping the cuts just shy of the layout lines. The 45° transition was later cut with a handsaw, and finally all of the inner edges were softened with a ¼-in. rounding-over bit.

Using his double-spindle horizontal boring machine, Whipple bored the rails and stiles for ½-in. by 2-in. dowels (two in each joint). After the doors were glued up, he hauled them all into town to a millworks where he had rented time on a large pneumatic drum sander. One pass on each face through the machine left the doors ready to finish.

Nowhere else in the house is the visual effect of thoughtful layering and simple, unmi-

Door casing and detail at head

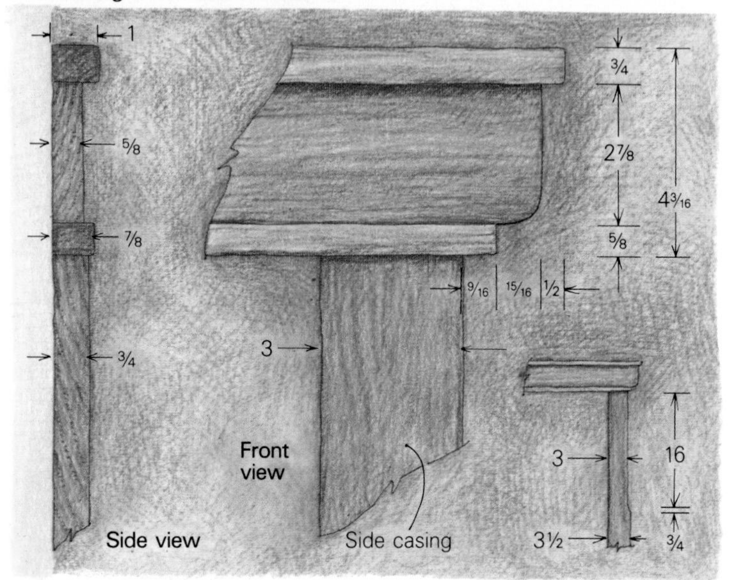

Kitchen cabinets and detail

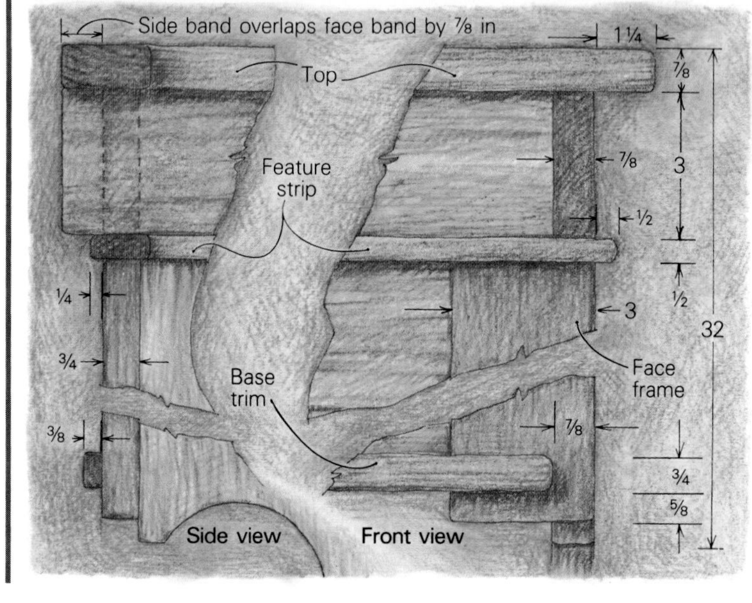

tered joinery more evident than in the door and window casings. As shown in the photo and drawing above left, the door casings (window casings are fundamentally the same) are made up of two vertical or jamb casings and a head casing. The jamb casings are 3½ in. wide for most of their length. About 16 in. below the top they jog inward ½ in. and narrow to a width of 3 in. This gives the lower parts of the casing the visual mass to balance the overhanging head casing, which recalls the yoke motif that's so common in Oriental architecture and furniture design. The design is disciplined without being severe; the construction is easy.

The stacked head casing happens in three vertical planes and in three horizontal planes. It's assembled from three separate pieces of trim, each one a different thickness and width, and each one a different length. From bottom to top of the casing, the sections of trim get progressively longer.

The lower trim strip (⅞ in. by ⅝ in.) is nailed to the main band (⅝ in. by 4 3/16 in.), which is radiused on its lower bottom corners. A top strip (1 in. by ¾ in.) is also nailed to the main band so that the whole head casing can be applied as a unit once the jamb casings have been nailed up. The resulting overhang appears to lift the ends of the head casing and produces a portal that looks a little like a *torii* gate.

Kitchen cabinets—The kitchen cabinets (photo and drawing above right) get their Craftsman look from the 3-in. wide face frames and from the way the upper part of the case is made from carefully proportioned pieces of mahogany trim. After assembling the main case in a conventional way, Whipple applied a feature strip to the top edges of the cabinets and surmounted that with a band whose corners overlap where they butt together. Then the top of the cabinet was sized to overlap the band by 1¼ in. Along with the frame-and-panel doors, the stacked trim enlivens the cabinetry and engages the eye—in a way that avoids the cute sentimentality that afflicts so many "designer" kitchens. □

Good and Small

A small floor plan needn't sacrifice elegance, privacy or entertainment space

by Sallie Hood and Ron Sakal

Animals and children have an uncanny ability to create for themselves places of just the size and character to provide a sense of well-being. These places are invariably small. Yet most adults seem to want big houses and large rooms. Not surprisingly, this "larger-is-better" philosophy has led to house designs that are spacious and imposing, but not at all cozy.

As architects, we genuinely enjoy small spaces. Fortunately for us, many of our residential clients have modest budgets, so we've had the opportunity to explore small-house designs. Sometimes we have to convince the client that a small house can still provide openness and privacy, despite its diminutive footprint. The key to making a small house work is to maximize quality in design and in construction. This is what we were able to do with the house shown here. It's the largest one we have built, with just under 1,400 sq. ft. spread over 1½ stories.

A lakeshore site—The house is located about 700 ft. from Lake Michigan in the small town of Union Pier, Mich., about an hour's drive northeast of Chicago. This shorefront area has a rich history of Craftsman-style houses. Narrow roads wind their way through the hilly sand-dune topography, and there are scores of small bungalows and summer cottages that predate the Depression. The woodwork is simple but well done, and terra-cotta flue tiles protrude from chimneys made with large pebbles ground smooth and round in the waves. Despite the occasional $500,000 mansion of more recent vintage, this is a good place to build a small house.

Elise and Alex Gorun contacted us in the summer of 1984, after having found a heavily wooded half-acre lot close to the water. The Goruns live and work in Chicago, and they wanted us to design a year-round weekend retreat with a full house's worth of rooms in just 1,200 sq. ft. (the minimum allowed by zoning for a 1½-story house). The house would need to accommodate Alex and Elise and their three-year-old daughter Amy, with extra room for occasional visits from the rest of the family: Christine, who is now in college, and Elise's mother and 96-year-old grandfather.

The Goruns had seen several of our renovation projects where we had had to make much of limited space, so they were confident that we'd be able to develop a workable design. Alex liked the local Craftsman-style vernacular because it reminds him of the old houses in Rumania, where he grew up. Elise wanted privacy, as well as space to entertain.

Making small seem large—Instead of an open plan with few rooms, this house is a collection of many different small spaces (see the floor plans on p. 31). There's a bathroom and two bedrooms upstairs. The master bedroom has an adjacent sewing room for Elise. Downstairs, there's a second bathroom, as well as a guest-room/sitting room, a living room, a kitchen, a dining room and an entry vestibule. The house also has a partial basement for storage, the furnace and a washer/dryer.

The final design did involve some compromises. We had to omit a master bathroom, trading it for a bedroom fireplace—a more romantic alternative. There's no breakfast area in the kitchen, but there is a convenient pass-through in the wall that separates the kitchen and the dining room. Also, we had to eliminate the garage, which in our initial design had a studio above it (both appear in the drawing on the previous page).

Decks, balconies and porches are valuable small-house features for several reasons. They're not usually included in the calculated square footage, and they're not exceptionally costly. What they do is give a much more open feeling to the space or spaces that they connect to. Upstairs, the Gorun house has three balconies—one is off the master bedroom's sewing alcove; another overlooks the front door. These balconies are small outdoor spaces, connected to the interior by French doors and sheltered by the overhangs of gable dormers (photos above and facing page). The third upstairs balcony is indoors. It overlooks the living room and is a play area for Amy.

The front porch (photo above) expands the entry significantly. The entry vestibule is small, but doesn't seem so because of the porch. The Goruns like to entertain, and this is a place where guests often linger before departing.

The back terrace is roughly 13x16. It fills the space between the two wings of the house, connecting the living room with the dining room and the kitchen. All three openings onto this terrace are through French doors. These were slightly more expensive than standard sliding-glass units, but they're much more at home in this Craftsman-style house.

The low parapet wall that separates the deck from the sandy backyard is shingled and trimmed out just like the house. This visual and structural link with the house is important. It integrates the terrace with the interior space and thus makes the house seem larger. The wall also functions as a bench. From this elongated seat, you're at eye level with anyone in the yard, so conversations can easily carry beyond the deck. Just off the terrace, against the kitchen's exterior wall, there's an outdoor shower, complete with colored tiles and a trellis enclosure.

Open up any catalog of typical house plans, turn to the section on 1½-story houses, and you'll find that this sort of program can't be had for under 2,000 sq. ft.—and that's without the fireplaces, balconies and terraces. Multiply the construction cost per square foot by the extra space, and you'll see why building small makes sense. At one point, the Goruns asked us to add a 2-ft. slice on both sides of the house. Even this proved too costly, and that's when we made the revisions that resulted in the final design. And as Alex points out, they've had to buy less furniture.

Overleaf: **Hood and Sakal's drawing of the house in winter includes a garage, which remains to be built. In the background, Lake Michigan stretches toward the horizon.**

From *Fine Homebuilding* magazine (Spring 1987) 38:35-39

Modern approach, traditional results—A friend of ours describes us as wolves in sheeps' clothing because we're really modern architects who trick people into believing that we've given them a traditional house. Our houses *do* look traditional, but we never start out with a preconceived notion of what a plan will look like. While we heed vernacular styles, we let the project's functional requirements design the house for us, and in this sense we *are* modern architects.

In this design, all the requirements dictated that the house be L-shaped and oriented so that the leg incorporating the entry would be parallel to the main road. The space created by the juncture of the two legs faces south and east, forming the terrace. Elise and Alex wanted the kitchen and the master bedroom to have east light. They wanted the kitchen, dining room and living room to be adjacent to the terrace. Everything else just fell into place. Our first tiny bubble diagram, drawn on a napkin in a pub, had all these elements in their logical places.

So, you see, modern architecture *does* design itself—but only through the bubble-diagram stage. After that, some formal rules of old-fashioned architecture have to take over. Every element in the design has to make sense in terms of all other elements. Things have to line up or somehow relate to each other. Inaccurately placed windows or doors, inconsistent trim details, elevations designed without regard for

Walking around. **From the street, facing page, the house presents a modest facade, with a small gabled balcony above an entrance porch. Wide eaves and gable overhangs form a protective brow that shelters the house from summer sun and winter storms. In the back, top, the design stretches up and out. Through small windows and another upstairs balcony, the second story overlooks a terrace that connects both wings of the house. On the north side, above, many small windows do the work of several large ones, with quainter effect.**

their relation to the interior design—or vice versa—and other design errors or oversights can prevent a small house from being a beautiful one. In a small space, your eye sees many details at once, and you can quickly compare them to each other. In a large space, you've got to look around and memorize various details for comparison, so you can be more forgiving.

Design takes a lot of time. Human bodies are bilaterally symmetrical, so human beings are accustomed to understanding that organizational system. And it's a good place to begin. But rigid classicism seems contrived for residential projects—especially small ones. Little houses don't yield to unrelenting bilateral symmetry because when you reduce each element to its minimal size, you just never end up with an even number of equally sized spaces. If you do chance to discover two spaces that are exactly the same size, the odds are slim that they are of the same relative importance. So you will always have an asymmetrical plan unless you force what need to be large spaces into small ones, or you expand small areas needlessly.

Each room, and each space that needs to be discrete within a larger room, must be well composed. To be truly comfortable, a room must tell its occupants where its center, edges, entrance, views and important features are. More often than not, some degree of bilateral symmetry is necessary to accomplish this.

By looking carefully at house elevations and floor plans, it is possible to see how bilateral symmetry is used, both in simple and in asymmetrical arrangements, to organize spaces and exterior elevations and to give visual clues about the functioning and the relative importance of various parts of the house.

In the dining room, for example, the two small windows and bookshelves on either side of the fireplace are the simplest sort of bilaterally symmetrical arrangement (see the floor plan, facing page). With the centerline of the French doors, they create a strong north-south axis that continues across the terrace and down the terrace stairs. The east-west axis of the dining room is formed in a more complex way. From the entry hall, you enter the dining room through an opening centered on that space. The east-west axis of the dining room appears to end at the solid wall behind the refrigerator; that wall effectively screens views from the dining room into the kitchen. Actually, the east-west axis continues through the refrigerator and ends at the kitchen sink. The actual centerline of the kitchen is located at the centerline of the seven little windows above the sink, and it coincides with the centerline of the outdoor shower and trellis. In the kitchen, both functional center and actual center are important, and they can coexist happily.

In the exterior elevations, similar principles apply. The west, or entry elevation, is at first glance a simple bilaterally symmetrical composition—a good clue that this is the formal, "public" side of the house (photo p. 28). However, a closer look at this facade reveals that the wall to the right of the entry has four windows, while the one on the other side has only three windows. This tells us that the room behind the four windows (the living room) is more important than the room with three windows (the study/guest room). Also, the entry door is flanked by only one window on its right—a coat closet prevents the placement of a window on the left. Thus the entry facade is saved from a rigid symmetry that would belie not only the modesty of the house but also the arrangement of rooms behind the facade.

Siting—Although southwestern Michigan is mostly farms and small country towns, this particular development was laid out like a suburban tract. The neighboring homes include some $500,000 builder houses on the lake shore as well as some smaller ranch-style houses nearby. Many of the neighbors have removed their trees to create lawns, but Elise and Alex wanted to be in the woods. We placed their house on the northern edge of the lot in a small clearing that required the removal of only one small tree.

The lot is essentially flat, on sandy soil just past the edge of the dunes. In the winter, Lake Michigan is visible from all the second-floor rooms. The lake shore gets much more snow than Chicago—sometimes 18 in. can fall in a few hours, so roofs must support heavy snow loads. During the summer, hot, humid days aren't uncommon. The area is always damp, and musty indoor conditions can get worse if a house is shut up during the week. Even though the house has central heating and air conditioning, we used large roof overhangs to minimize the need for them. These 3-ft. overhangs keep sun out in the summer and let it in during the winter. They also make it possible to keep windows open on hot, muggy, rainy days.

A complex roof and a relatively large number of windows give a small house greater presence. A 1,200-sq. ft. two-story house with a simple roof would have been less expensive than this 1½-story house with its seven gables, but local zoning required two-story houses to be a at least 2,000 sq. ft.—more area than the Goruns wanted to pay for. And a simple roof would have lacked vitality and substance. We could have used far fewer windows to comply with minimum requirements for adequate lighting and ventilation, but minimal windows would have seemed meager. Windows needn't be large—several small windows are often more effective than one large one, and they frame views more dramatically. Energy efficiency needn't be sacrificed either, but each room must have views in as many directions as possible, particularly when the inhabitants want to feel surrounded by trees.

Under construction—Though we wanted a Craftsman-style appearance and feel, we knew that the budget wouldn't allow for many custom items. We'd have to use off-the-shelf materials and standard construction details to keep costs under control. We try to design what we might build ourselves. Then we let the pros do it better than we can.

The general contractor, Dave Prusa, is a local builder who learned the trade from his father. To Dave and his crew (Rudy Pruse, Rick Oney and Dennis Granke), our design presented many construction details that they hadn't tried before. The roof framing was the most challenging part of the job, but there were also balconies, French doors, unusually broad soffits and small windows everywhere. Fortunately for us and for the Goruns, these fellows did an excellent job on the construction and on the detailing.

The framing is ordinary platform style, using 2x6 studs, 2x8 and 2x10 floor joists and 2x8 rafters. The dormer walls are continuous with the first-floor walls, so these 1½-story sections were framed all at once.

At the gables, the outermost rafters are double 2x8s braced solidly against the end wall by a ladder of short 2x8 lookouts. These end rafters are tied at their peaks into the cantilevered ridge beam. Down below, they rest on a double 2x12 beam that is cantilevered 3 ft. beyond the wall and bolted to the wall's double top plate. Thus supported, the gable overhang should be able to stand up to snow and wind loading during the most severe winter.

We used ventilated aluminum soffits on the undersides of all eaves and gables. Apart from promoting excellent ventilation (along with a continuous ridge vent), this material has good texture and scale—it is our late-20th-century answer to unaffordable beaded fir soffit boards. The large outdoor terrace is made of pressure-treated pine. It has a 1x6 border and a "rug" of 1x4s to better define the outdoor "room."

Elise and Alex wanted their house to be both charming and elegant, and they also wanted it to be easily maintained. Alex insisted on using vinyl-clad windows to eliminate the worry of repainting window trim. We chose Andersen because we like their standard sizes and because they always give us a good price.

The Goruns wanted blue/grey-stained shingle siding. A wood shingle roof proved too expensive, but a roof of thick-profile, self-sealing asphalt/fiberglass roof shingles (Timberline brand by GAF, Building Materials Group, 140 West 51st St., New York, N. Y. 10020) was an acceptable substitute. Elise wanted lots of French doors, and Alex asked us to incorporate two old fan windows he owned. These elements in combination make the house seem a little more like East Coast Shingle-style houses than like the Prairie-school-influenced Craftsman houses. Brown shingles and custom-made Craftsman-type doors (prohibitively expensive) would have made the house more truly regional. But this house is one of the few new residences that has anything to do with the fine cottages and farmhouses that abound in this part of Michigan.

Designing small houses is a lot like designing a Chinese puzzle—it's time-consuming because so many crucial parts need to fit into limited space. But after having taken this approach with a number of small houses, we're convinced that it's a good way to proceed, even if you're designing a 4,000-sq. ft. residence. The only successful large houses we've seen are those where alcoves, bays, recesses, compartments and niches are built within rooms until the scale is finally reduced to a human one. So why not design a house that is scaled in a friendly and habitable fashion to begin with and spend all that other money where it counts—on high-quality materials and good workmanship? □

In the living room, above, quarter-round windows flank the chimney in the house's gable end. This fireplace is one of three; the other two are in the dining room and the master bedroom. At left, a dining-room view of the kitchen shows a ribbon of small windows. Upstairs (above left), the master bedroom has a small annex with French doors that open onto a gabled balcony at the back of the house.

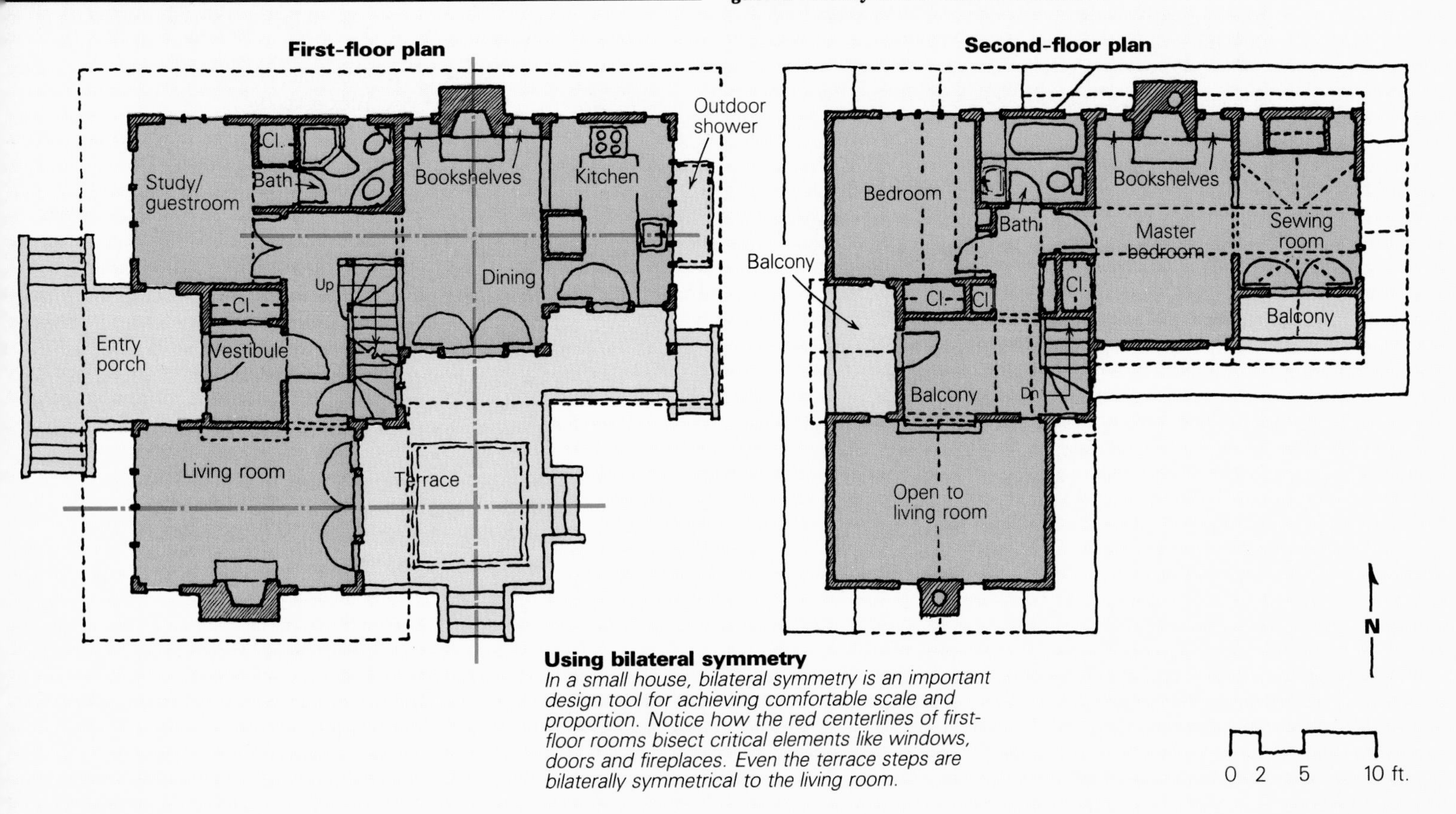

Using bilateral symmetry

In a small house, bilateral symmetry is an important design tool for achieving comfortable scale and proportion. Notice how the red centerlines of first-floor rooms bisect critical elements like windows, doors and fireplaces. Even the terrace steps are bilaterally symmetrical to the living room.

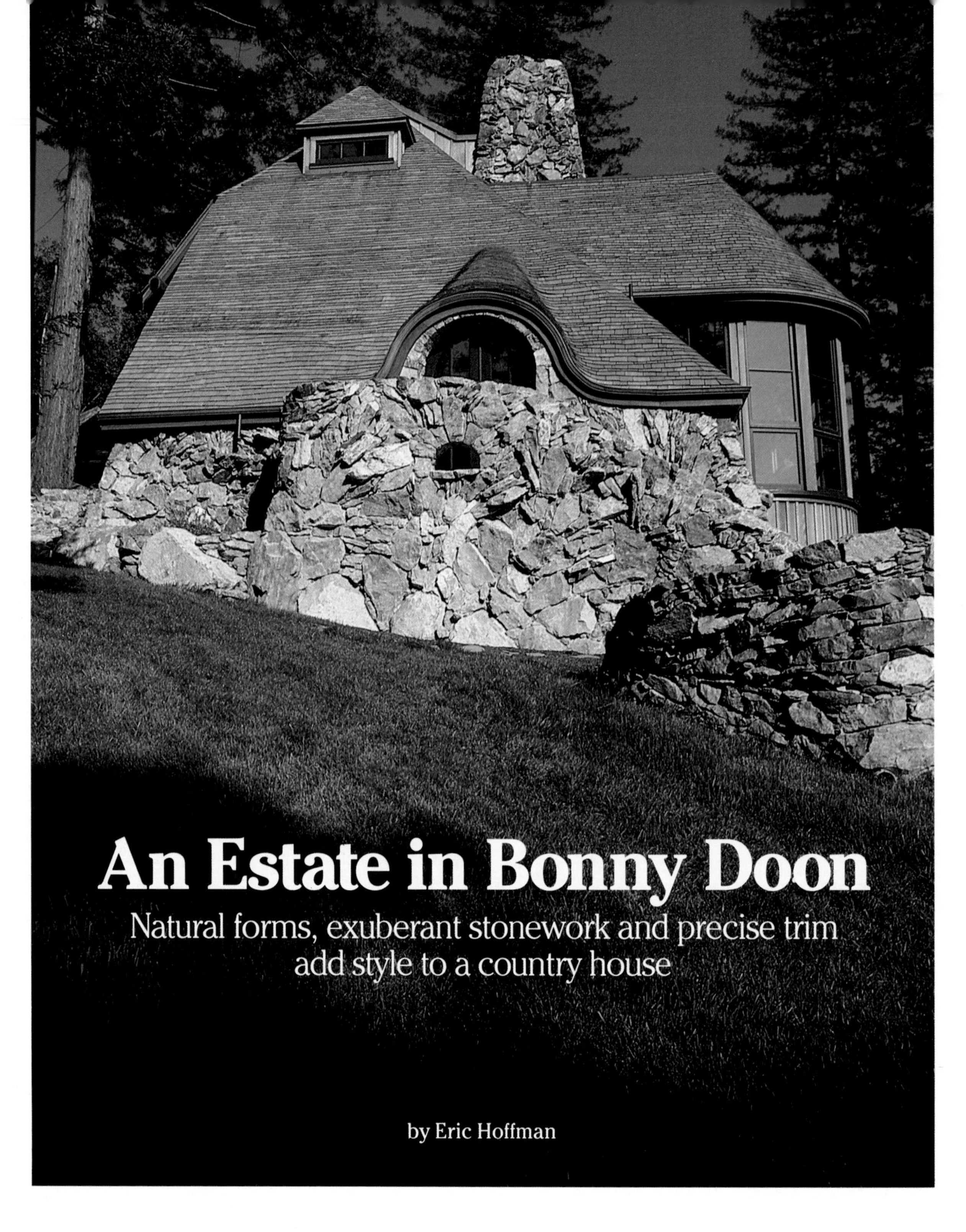

An Estate in Bonny Doon

Natural forms, exuberant stonework and precise trim add style to a country house

by Eric Hoffman

It's about an 80-mile drive from San Francisco, south on Highway 1, to the turnoff to Bill Cunningham's dream. The major landmark is the cement factory in Davenport, a little seaside community that started out as a whaling station. The road to Bonny Doon intersects the highway a couple of hundred yards past the town, then winds up the wooded contours of a small canyon to the first escarpment of the Santa Cruz Mountains, 1,000 ft. above the sea. The ridge top is part meadowland and part forest, with stands of mature redwood, oak and fir trees.

A visit to a 270-acre parcel of land that was for sale in the middle of this setting convinced Bill Cunningham that it was time for a major career change. He had a successful dentistry practice in nearby Santa Cruz, but that wasn't the creative outlet he wanted anymore. Instead, he saw the rural landscape on the ridge as a chance to involve himself in his long-standing appreciation of Arts and Crafts architecture, especially as practiced by the likes of the Greene brothers and Bernard Maybeck. Cunningham's wife Robin has a passion for plants, so her landscaping contributions would be a welcome addition to a residential development in these woods. The two had visions of a community of finely wrought homes on five-acre parcels, with a large tract of land set aside as open space.

And so with no experience in the construction business and no architectural training, Cunningham sold his dental practice and put up the proceeds along with his life savings to become a real-estate developer. He found enough investors to outbid a lumber company for the land, and set about to build a house on speculation that would establish the tone for the homes he hoped would follow.

A suitable style—Cunningham wanted to use natural materials extensively in the first house, with colors and forms that would make it seem a part of its setting. As he inquired among friends and associates for residential designers, the name of Clarke Shultes came up. Shultes lives in Santa Cruz, where he runs an office specializing in eclectic houses with a slant toward the Arts and Crafts period. The two men hit it off, and Shultes was hired to design a house that would be appropriate for the wooded setting.

Shultes' homes are usually built in established

From *Fine Homebuilding* magazine (August 1987) 41:66-69

neighborhoods, and his designs sometimes fall victim to neighborhood pressures to conform to recognized styles. But on the ridge top in Bonny Doon he had free rein to design a house from scratch. For inspiration, he looked to the trees. His goal was to invoke the shapes, colors and textures of the surrounding forest in the form and finish of the house.

Wherever he could, Shultes worked peeled logs into the structure—the entry is flanked by log columns, and the deck railing is supported by short log posts. The exterior siding is redwood board and batten. Inside the house, log beams are used as accents in the living spaces. On the west side of the house, the rounded forms of an eyebrow dormer and a two-story turret repeat the cylindrical tree-trunk forms. The steep roofs (16-in-12) mimic the shapes of the trees (photo top right). Wavy courses of cedar shingles cover the roofs, recalling the irregular patterns and soft shapes of coniferous foliage. Instead of sharp hips, the roof planes intersect with soft radii.

The price in time and effort to render these ideas was considerable. To fabricate the roof decking at the rounded hips, builder Greg Howerton's crew had to use built-up layers of ¼-in. plywood, pulled into place with a come-along and screwed to the framing. Then the roofing crew had to wilt the shingles with boiling water to coax them into following the contours of the roof. To wed the roofs to the forest, Shultes had the shingles sprayed with Copper Green, a product normally used for its preservative qualities. Says Shultes, "When the fog rolls in and your vision is somewhat blurred, the house fades into its surroundings because of its shape and color."

While the building's roof forms and colors unite it with the forest, it is the stonework that anchors the house to the site.

Clarke Shultes' design for a house in a redwood forest pays homage to the trees in form and color. The steep roofs with rounded hips echo the silhouettes of the trees in the background, and the cedar shingles tinted green mimic the surrounding foliage.

Above, Hansel Kern taps a rock in place along the base of a low wall built against wood framing. In California, stonework has to be reinforced to resist earthquake jolts, so Kern has attached masonry ties to the studs. The ties in turn are folded around ¼-in. pencil iron reinforcing steel, which can be easily bent around the irregular rocks. Kern's work is characterized by invisible mortar lines (facing page) and stones carefully juxtaposed to amplify their differences in shape and size.

Stonework—Before he packed up his tools for the last time on the Cunningham job, stonemason Hansel Kern had built two immense fireplaces, several intricate stone walkways, numerous walls and a true arch. Counting the 200-yd. retaining wall along the entry road, Kern grouted and stacked over 1,200 tons of green serpentine stone that Cunningham had hauled 180 miles from the Sierra Nevada Mountains.

Hansel Kern is the son of the late Ken Kern, the author who inspired a generation of back-to-the-land builders with his books on alternative construction techniques. Hansel Kern has been making stone walls since he was 10 years old. A dozen years ago, when he was 16, he started crisscrossing California, busily laying rock for everyone from "aging hippies who need a stone fence to keep their goats from wandering away to people like Bill Cunningham who realize the art value of rock work."

Over the years as a stonemason, Kern has worked with dozens of varieties of rock and has settled on three favorites: green serpentine, desert fieldstone, and tailings from Sierra gold mines and hydroelectric shafts. Kern likes these materials best because they have sharp angles that will fit together with a minimum of mortar. He prefers not to shape the rocks into conformity, and he often works an entire day without having to chip away at a stone to get it to fit into its niche.

Kern refuses to work with river rocks because it's a lot like stacking blocks, which he says is boring. Says Kern, "Give me a couple of dump-truck loads of green serpentine straight from a quarry and I'll be able to do something special."

Because California's earthquake-conscious building codes forbid unreinforced masonry, Kern's residential stonework is done as veneer. If he's working on a wood structure, he attaches masonry ties to the framing on 16-in. centers, spaced about 10 in. apart horizontally (photo above). California building codes call for a masonry tie for every 2 sq. ft. of surface area for veneer-type masonry, so Kern's reinforcing technique amounts to nearly twice as many ties as are normally required.

If he's going to work along a poured-concrete wall, Kern makes sure ties are part of the form-building. Standard masonry ties, which are 8 in. by ¾ in. by $\frac{1}{32}$ in., have to be shoved through precut slots in the foundation forms and wired to rebar before the pour. With block walls, the ties have to be worked into the mortar at appropriate intervals. Each tie is bent into an L-shape, and the foot of the L is hooked over the inside of the block so it can't readily be pulled out. In situations where ties have not been provided in either block or concrete surfaces, Kern simply nails the required number of ties to the wall with a powder-actuated fastener.

A layer of stone veneer is very heavy, so Kern

Windows and door are cased with rounded trim stock milled from select 2x4s (left). Half-lap bird's-mouth joints at trim intersections (detail above) are used throughout the house, and were milled in place with routers.

makes sure that the building's foundation is substantial enough to carry its weight. Usually the builder will widen the footing before the foundation is poured to pick up the load of the stone. But sometimes this necessity is overlooked—even by the inspectors—and a separate footing for the stonework has to be grafted onto the building's foundation.

Kern folds the ties around horizontal lengths of ¼-in. pencil iron so that the iron ends up about 3½ in. from the wall. He prefers pencil iron to ⅜-in. rebar because the ¼-in. stuff is flexible enough to bend easily around the irregular shapes of the rock as the stones are mortared into place.

Once underway, Kern's style is all his own (photo p. 32). He sometimes uses one kind of stone as a main course and another to create veins of a different tone. He often develops contrasts by placing smooth stones next to jagged rocks, or juxtaposing small pieces next to hefty boulders. Sometimes Kern will plant a long stone so that it points almost straight up, beginning a line of subsequent stones that seem to burst geyserlike from the earth. In other places, the stones repose in sedimentary layers. One of his primary rules is to "balance" a wall. "Too often masons put all the big stuff at the bottom and the little rock at the top as if the top is where all the leftover material belongs," says Kern. "I like to put large stones and small stones throughout a wall. Big stones high in a wall balance a work and create curiosity."

Mortised trim—To maintain the continuity of the rounded look on the interior, Shultes drew up a trim detail that carries on the radiused feel of the peeled logs, but in a very precise manner. The detail uses hefty casings fashioned from kiln-dried vertical-grain Douglas fir 2x4s. Where trim members intersect, they are joined by way of a V-shaped half-lap on the end of one member and a V-shaped mortise in the other (photos left and above). The detail looked good on paper, but making it would present some challenges to the trim carpenters. The casings would eventually be finished with lacquer, so there wouldn't be any room for putty in these joints.

To make the trim efficiently, Cunningham turned to a pair of woodworkers who are well known in Santa Cruz for their work on sailboats. At Cunningham's suggestion, Rick Bornhurst and Yarrow Smith moved their cabinet shop into a renovated barn near the house, and pondered workable solutions to making the trim and the fastidious interlocking joinery shown in Shultes' intricate elevation drawings.

Milling the trim was straightforward work. A section of the trim shows the rounded profile to be half of an ellipse. The woodworkers ordered a custom-made cutterhead for their shaper to match the profile drawn by Shultes, and then ran 3,000 lineal feet of 2x4 stock through the machine. Each piece got two passes per edge. The first took most of the unwanted material, and a second light touch cleaned up the surface.

Shultes wanted the trim members to be full-length pieces, unbroken by scarf joints. In some places, as many as five vertical pieces intersect a horizontal member. Some verticals run between two horizontal pieces, and their length has to be on the money. Recalls Bornhurst, "In essence, hundreds of trim pieces became an interlocking matrix, and if a piece didn't fit perfectly, other pieces would be affected." Bornhurst and Smith reasoned that the most accurate and efficient way to cut the many joints would be with template-guided routers. The templates would have to clamp to a piece of installed trim, so that the mortises could be cut in place.

Templates and bits—To make the templates, the two woodworkers had first to think about the bits they were going to use to make the cuts. They decided on straight-flute bits with bearings mounted above the cutters to follow the contours of the templates.

For the mortise cut they used a ¼-in. straight bit with a ⅝-in. bearing, as shown in the drawing at top right. They used the small-diameter bit for the mortise because they wanted as small a radius as possible at the point of the V. To complicate matters, the mortise template had to be oversized to allow for the difference in diameter between the bit and the bearing. In this case, the offset amounted to 3/16 in. per side. Once they had a master template made out of ¼-in. Plexiglas they made a dozen copies using a router with a flush-trim bit to duplicate the master. They knew the templates would be subject to gouges from accidental router cuts, and once a template is damaged, it's tough to repair it.

Making the half-lap, or beak, portion of a joint took three steps. First, the end was rough cut with a power miter saw. Next, a male template that corresponded to the profile of the V-shaped mortise was clamped about ¼ in. from the rough cut, and the finished edge cleaned up with a ⅝-in. dia. flush-cut trim bit (drawing, middle right). Finally, the template was removed and the trim stock turned over. At this point the other end of the template could come into play. Its straight side was clamped to the backside of the beak, and the wood was routed out to complete the half-lap joint (drawing bottom right). By clamping the templates directly over carefully scribed layout lines, the trim carpenters could work to within 1/64 in.

When they first started experimenting with the templates, Bornhurst and Smith bought some bits from a supplier with the bearings already in place. The bits were expensive, and after a predictable number of cuts the carbide edges wore down and the bearings failed. Faced with hundreds of joints to cut, they decided to try to reduce their tool expenses by adding bearings to ordinary straight-flute bits. To do so, they affixed router-bit bearings from their local supplier to the ¼-in. bit shafts with Loc-tite, a glue normally used to keep bolts and nuts from separating on machinery that is subject to intense vibration. Smith and Bornhurst found that their homemade versions lasted just as long as the more expensive ones, and that the bearings would give out long before the glue bond failed. □

Eric Hoffman writes about houses and travel from his home in Santa Cruz, Calif.

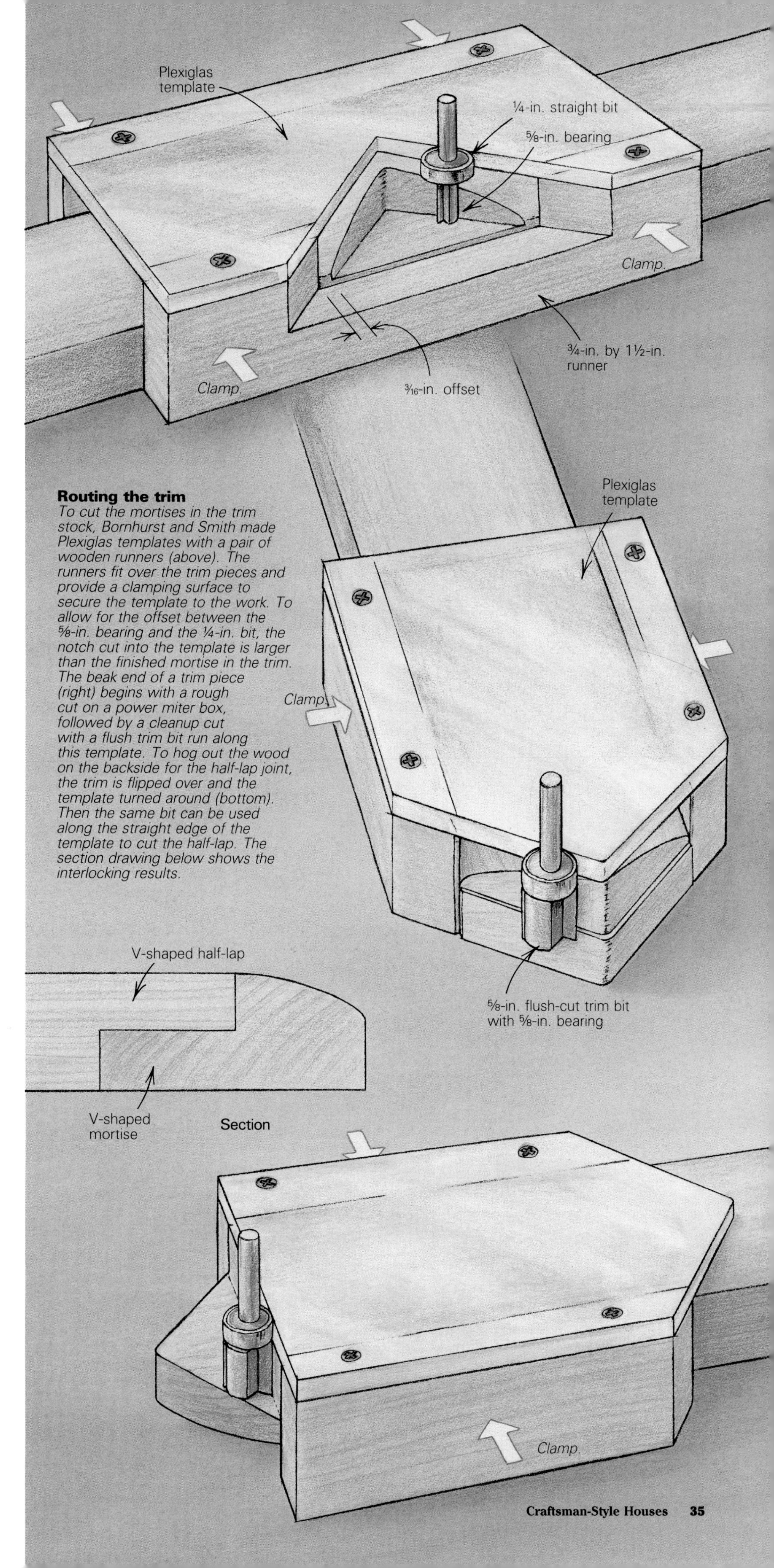

Routing the trim

To cut the mortises in the trim stock, Bornhurst and Smith made Plexiglas templates with a pair of wooden runners (above). The runners fit over the trim pieces and provide a clamping surface to secure the template to the work. To allow for the offset between the ⅝-in. bearing and the ¼-in. bit, the notch cut into the template is larger than the finished mortise in the trim. The beak end of a trim piece (right) begins with a rough cut on a power miter box, followed by a cleanup cut with a flush trim bit run along this template. To hog out the wood on the backside for the half-lap joint, the trim is flipped over and the template turned around (bottom). Then the same bit can be used along the straight edge of the template to cut the half-lap. The section drawing below shows the interlocking results.

A Craftsman-Style Renovation

Long Island architect turns a boxy cottage into a bungalow

by Eric Woodward

We didn't start by looking for a Craftsman-style home. In fact, our first interest was in Mission-style furniture from the Arts and Crafts period. My wife Hilary started our collection in 1980 with two chairs, found at a local antique store. Although we were attracted to the sturdy, simple design of the furniture, we didn't immediately make the connection that perhaps our new house should be in the same style. After many frustrating sessions traipsing around town with various real-estate agents, we found our house through the real-estate version of video dating. It was love at first sight.

The lot, in Southampton, N. Y., has a meadow with trees to the south of the house and backs up to a large wooded area. The house was built in 1928 as the playhouse for an estate across the street. Originally, the interior was one open room 30 ft. wide and 40 ft. long. Over the years, little was touched on the exterior, but inside five small drywalled rooms were built. In the 1950s, the owners had added a concrete-block bathroom and one-car garage against the north side of the house, blocking the view into the woods.

Though the house attracted us, as an architect I knew that a major renovation lay ahead. A new house could be done just as cheaply, but here was an opportunity to work in the discipline of the Craftsman style and to live with something partly antique. My goal was to integrate our family's lifestyle and the house's architectural style.

The work begins—Hilary and I completely tore out the interior partitions in the house, clean down to the last projecting nail head. To be saved were the first-floor casement windows, the building frame and sheathing, the fir floor and much of the chimney. Everything else would be new.

The garage and the concrete-block bath were removed from the north side of the house. This opened up the view and gave us access to start digging out underneath the house for a new concrete-block foundation. The house was originally built on locust posts, so it was relatively easy to excavate beneath the building. We hired a specialist in digouts, who erected temporary posts to support the house while the block foundation was completed. We had a party when all the excavation work was done, and I remember how the house shook on its posts as children ran around. Luckily, everything stayed in place.

New plans—By this time, I had drawn the basic set of working drawings—three sheets of plans and elevations. By the end of the job I'd completed 20 sheets of details and special conditions. Most of this extra design work related to Craftsman-style details that I'll discuss later.

In the basement, which opens out to grade, we have a sewing room, a utility room, a darkroom, a changing room with a small sauna and an office. All these separate spaces radiate from a central area we call the rumpus room. To make room for the office, I extended the foundation out 8 ft. beyond the walls above. This space has a low-sloped roof (with skylights) that doesn't intrude on the otherwise traditional exterior.

On the second floor, we planned and built a master bedroom, a dressing room, a bathroom and a nursery for our soon-to-be-born daughter. We created more headroom upstairs and brought in plenty of natural light by adding two shed dormers. These have new divided-lite casement windows similar in style to the original casements on the main floor. Outside, the projecting rafter tails of the dormer roofs match those of the original rafters—a standard Craftsman detail.

Planning the main floor was not as easy. There was no question that the large fireplace on the south end of the house would be the focus of the living room. We opened this space up to the

A modernized cottage. **Built in 1928 as a playhouse, this humble structure in Southampton, N. Y. (facing page, top left), was largely neglected until renovation began in 1985. It had little to offer beyond a rustic setting, a low price and a few Craftsman-style features. Rebuilt inside and out, the new house pays tribute to the Craftsman style without sacrificing comfort or convenience. The chimney's crumbling stucco veneer was replaced with a covering of round stone, flared at the top for a more graceful shape (facing page, top right). Two pairs of dormers straddle the ridge, bringing in light and upstairs headroom. New shingles were stained an earth tone, accented by red and green exterior trim. A fieldstone patio and stone retaining wall surround the pool (right). A cedar trellis leads to the front door.**

Photos, except where noted: Eric Woodward

Alan Becker

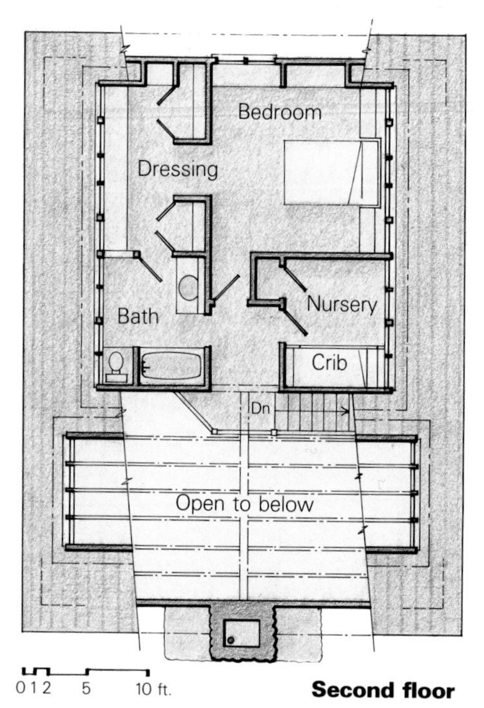

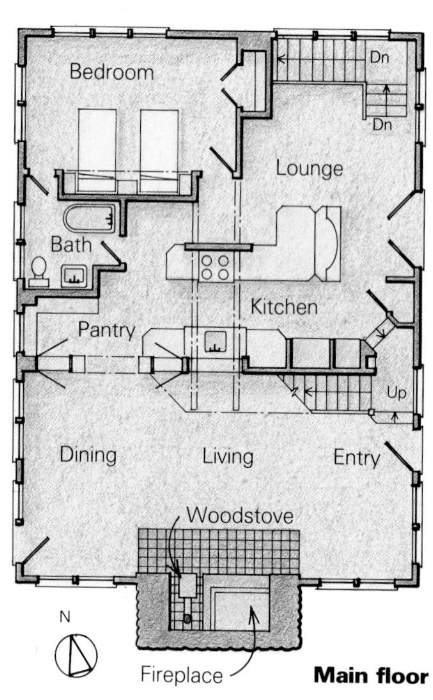

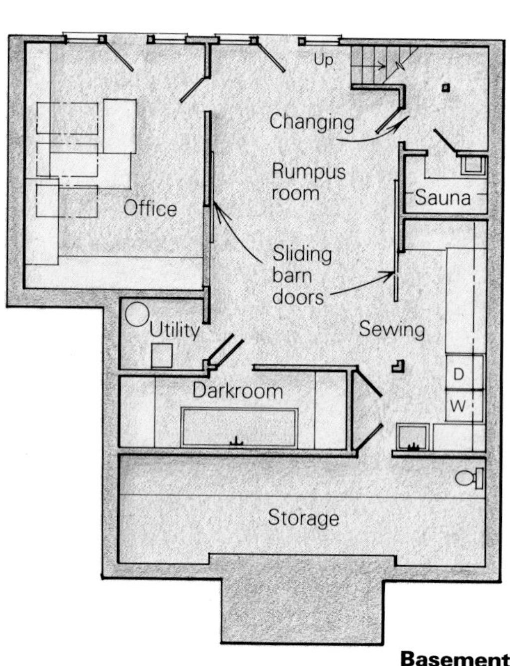

rafters, adding two more shed dormers. These bring light into the room through the existing rafters, which were left in place (bottom photo, facing page).

In addition to a bedroom and bath, we had planned to have a formal dining room with the kitchen at the far north side of the house. But the day before we were to start framing, Hilary said, "Wait a minute, this plan is all wrong, the dining table should be near the fireplace." Hours later we had completely juggled the plan, placing the kitchen at the center of the house (as shown in the middle drawing at left). In this central location, someone at the "control point" between the sink and stove can see to all parts of the first floor.

The dining area in the living room is separated from the kitchen by two built-in, double-sided, Craftsman-style cabinets. The placement of the stairs shields the kitchen from the entry. North of the kitchen counters is an eat-in bench and lounge where children can play or watch TV. A pantry, bathroom, closets and a guest bedroom fill the main floor's northeast corner.

Shingles, windows and Craftsman colors— Ken Rafter was our carpenter through all phases of the project, from framing to trim work, while I was the project coordinator. Construction proceeded in a normal sequence with a number of special projects relating to the old house. I contemplated reusing the existing 1928 asbestos-cement roof shingles; they appeared not to have weathered much. But the work involved in carefully removing and reapplying thousands of heavy shingles caused me to abandon this approach. After stripping the roof down to bare sheathing and completing the dormers, all roof surfaces were covered with red cedar shingles.

Ambitiously, Hilary and I hand-dipped each shingle in a mixture of clear Cuprinol and Cabot's Mission Brown and Evergreen stains. I even designed a drying rack that held seven bundles of shingles without lap marks—enough to keep the carpenters supplied for a full day. By the time the walls were shingled, we decided the stain could be painted on. We used a red stain, thinned considerably with clear preservative to give a very subtle tone.

Exterior window and trim colors are not so subtle. Here we were partly inspired by Karl Larsson, a turn-of-the century Swedish artist whose work celebrates country life. Bright colors are important in many of his paintings. The strong red and green trim colors we chose accent the Craftsman features with brilliance, and contrast nicely with the dark shingles.

Color selection inside required much experimentation. We often found ourselves saying "add more raw umber" to tone the color down toward the earth-based pallet typical of Craftsman style. Doors and sash on the main floor were painted a mustard color; walls are an earthy Mexican red. Ceilings are off-white. Upstairs, we used straight white for walls and ceilings. Trim is pale lavender, with muted green for the windows.

The original windows that we reused were scraped, sanded and reputtied before painting. In addition to putting a plastic spring-type weatherstrip around each of the inward-opening sash, we also added wood stops to hold brown aluminum-framed storm and screen panels. These panels don't interfere with views, and they can be removed from inside the window, simplifying seasonal changing.

Chimney and fireplace—Extensive cracks in the massive chimney caused some concern as we turned to this part of the renovation. The chimney had been built with a brick lining that

The stairway, with its Craftsman-style railing (facing page, top), leads to a small balcony overlooking the living and dining room. Dormers were added above existing rafters. The head casing for windows is continuous and works out flush with the top of the mantel, whose carved inscription reveals the name of the house and its construction dates. Leaving the original walk-in fireplace intact, Woodward built a raised hearth across two-thirds of the opening (right), allowing space for a woodstove with its own stainless-steel flue.

From *Fine Homebuilding* magazine (October 1987) 42:26-31

Photos this page: Alan Becker

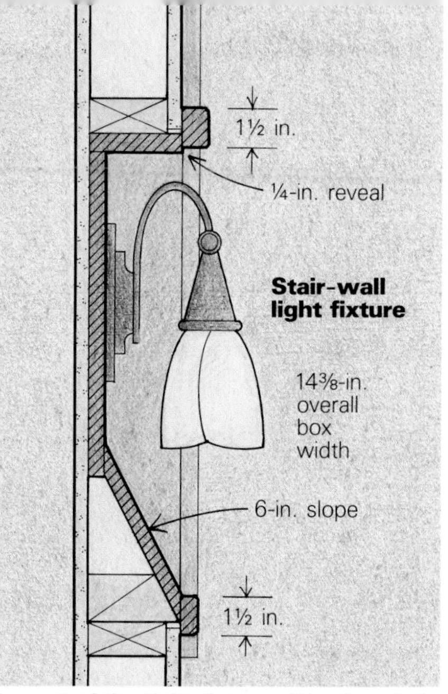

The largest of the light fixtures Woodward designed, of oak and frosted and colored glass, is above the dining-room table (below). The drawing above details a fixture above the stairs.

was covered by 8-in. concrete block finished with stucco. Fortunately, the lining looked sound, and the cracks were due to water penetration from the top rather than to footing failure. A demolition crew dismantled the chimney down to the roofline and also removed the 8-in. concrete block down to grade. Then a new veneer of round stone was laid up on all three exposed sides. Local mason Frank Mondello laid up the veneer. Working carefully, he flared the top of the chimney slightly, giving the mass a more graceful look. This new version seems more in keeping with Craftsman style than the old.

Inside, the fireplace was fairly sound. Some firebricks were repointed, and a new coat of stucco was applied over the old. The original walk-in fireplace was dramatic, but not very practical for daily use. We built a raised hearth across two-thirds of the firebox and installed a woodstove with its own stainless-steel flue in the other third. This gives us a choice between woodstove or fireplace heat without significantly altering the original hearth.

I designed a new Craftsman-style mantel that holds two old light fixtures from the house. Walter Kluge carved our name for the house, "Woodcroft," into the mantelpiece, along with two important dates: 1928 and 1985.

Interpreting the Craftsman style—Interior trim and woodwork required untold hours of design time. In working out these many details I wanted my designs to be true to the Craftsman style without being copies. To be avoided were the two-dimensional pastiche interpretations that are often seen in Post-Modern designs. I wished to honor the Arts and Crafts movement by doing new creative work as it might have been done originally.

The Arts and Crafts movement stressed collaboration between designer and craftsman; the designer must understand the nature of materials, while the craftsman's skill brings the beauty of the design to life. Simple embellishments, a love of craftsmanship and an appreciation of natural materials lend this style its characteristic warmth.

When interior trim is done in Craftsman style, flush connections are avoided. Instead, intersecting pieces are made different thicknesses and allowed to run past each other. Thus the character comes from the expressed nature of the connection rather than from applied decoration. This treatment was used frequently at Woodcroft—for casings around windows, doors and beams, and for the built-in cabinetwork throughout the house.

Another way that warmth is given to the design is through careful attention to scale. From English cottages and Dutch country houses, Craftsman designers borrowed the idea of creating a frieze above windows and doors. In our house, we extended the window and door head casing to form a continuous band around the first floor. As was typical, the ceiling color is brought down on the wall above the trim, giving this space a human scale despite the cathedral ceiling (photos previous page).

The wood we chose was clear southern yellow pine, partly for its lively figure but mostly because of its low cost. The carpenters were not happy because yellow pine is very hard to work, but our worst fears of extensive warping have not come true. As many woodworkers know, this wood doesn't take stain well. Staining an apparently clean piece of wood yields a surface that looks as if it had been beaten with a mallet.

The natural blond of the yellow pine did not go well with the rich russet on the walls and the mustard-colored frames of windows and doors. The solution on the trim was a coat of clear wax followed by a coat of tinted wax. I used Butcher's Bowling Alley paste wax (Butcher Polish Co., Marlboro, Mass. 01752) for both coats. I tinted the second coat myself, mixing about 1 teaspoon of standard liquid paint pigments (a combination of red, raw umber and yellow) with about 2 cups of wax. The wax was applied and buffed with soft cloths.

On the kitchen cabinets (built from yellow pine by cabinetmaker Tom Perez), I wanted a matching finish, but it had to be scrubbable. Wax alone wouldn't be good enough. After numerous experiments failed, I came up with an innovation—stained polyurethane. First I applied an even coat of slightly thinned satin polyurethane and sanded it very lightly when it had dried. Then I applied a mixture of turpentine and paint pigment (the same as went in the wax) to the dry polyurethane. This looks like wet gunk, but as you rub it off the turpentine dries, and with a little practice just the right amount of stain remains on the poly. At this point it is still reversible: if the finish is too dark or uneven, a turps-dampened cloth will remove the pigment. When the color is right, you seal it in with second and third coats of polyurethane. The stain doesn't come off on the brush. All the kitchen cabinets were done this way, and it is almost impossible to tell their finish from that of the waxed trim.

Being consistent to the period is relatively easy with plumbing. For the main-floor bathroom, we found an old claw-foot tub and a pedestal sink. Besco (729 Atlantic Ave. Boston, Mass. 02111) supplied us with some reconditioned faucets, and Renovator's Supply (4651 Renovator's Old Mill, Millers Falls, Mass. 01349) had a reproduction shower set for the tub. Hilary even consented to using an old kitchen sink when I found one with good enamel and two large integral drainboards.

In the kitchen, the countertops have a yellow pine edging around ¼-in. thick Corian infill (photo right). A recessed refrigerator and a wall-mounted oven/microwave keep the appliances from being too obtrusive, and the 20-in. restaurant range is suitably rugged looking.

Good Craftsman-style lighting fixtures are not so easy to make; reproduction fixtures are rare, and reasonably priced antiques are rarer still. Although the standard schoolhouse-type white globe is good in places, I wanted more sophistication. The chandelier over the dining table shown in the photo at the bottom of the previous page) received special attention. Its wood frame has predominant horizontal members extending past the "hanging" corner pieces, similar to the theme of the house trim. Glass is set into rectangular cutouts consistent with the stair balusters and our Limbert dining chair backs.

Looking at the details that we incorporated into our house, some people say "Frank Lloyd Wright," others say "Greene and Greene," while still others say "Japanese." This is no coincidence. Frank Lloyd Wright in his early years, the brothers Greene and Greene in California, and

The bed in the upstairs suite (facing page, top) is centered beneath a shed dormer and a low wall of built-in cabinets. Throughout the house, trim and cabinets are yellow pine. In the kitchen (right), below the continuous head casing, walls are painted an earthy red. The kitchen, at the center of the house, contains a salvaged double-drainboard sink. A window in the cabinetry allows a living-room view. Countertops are ¼-in. thick Corian, edged with yellow pine.

Gustav Stickley all drew some inspiration from Japanese woodworking.

The landscape also goes with the house. A small gunite swimming pool edged with boulders is set midway between the first-floor and basement levels (bottom photo, p. 37). Retaining walls are also boulders, but dry set. A fence for roses with a rustic twig gate separates the driveway from the yard. A cedar post arbor for grapes frames the entry walk. A small lawn, a flower garden and Hilary's raised vegetable beds fill our enclosed front yard. The shrubs we planted are a mixture of red twig dogwoods and broad-leaf evergreens, which pick up well on the colors of the house. □

Eric Woodward practices architecture in Southampton, N. Y.

The street side of the McGregor house, above, is flanked by a porch that gestures in a welcoming manner to the neighborhood. At the same time, there are few windows, thereby preserving the occupants' privacy. This house has a flat roof, allowing the horizontal quality of the design to assert itself. The house is primarily a post-and-beam structure that emphasizes the bones of the building for both style and organization. As the drawings below show, the bays between the posts in the north/south direction are 10 ft. wide, and areas for specific activities fall within these bays.

1 Porch
2 Entry hall
3 Powder room
4 Guest room
5 Family room
6 Living room
7 Open to below
8 Dining room
9 Kitchen
10 Breakfast nook
11 Deck
12 Mudroom
13 Bedroom
14 Recreation room
15 Laundry room
16 Bathroom
17 Garage
18 Storage room
19 Furnace room
20 Master bedroom
21 Closet
22 Master bath
23 Whirlpool
24 Den

Lower floor plan

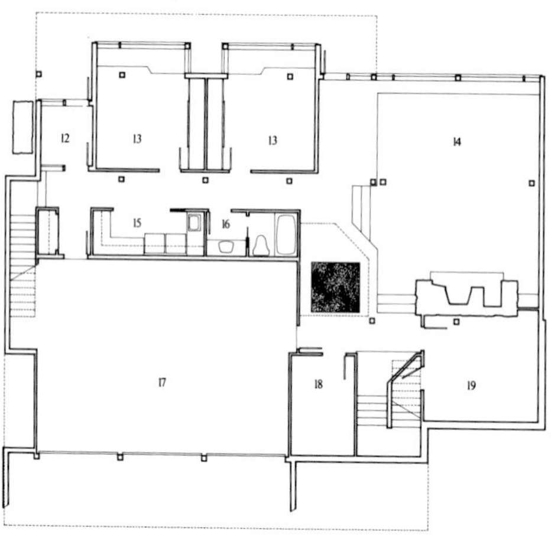

Main floor plan

Upper floor plan

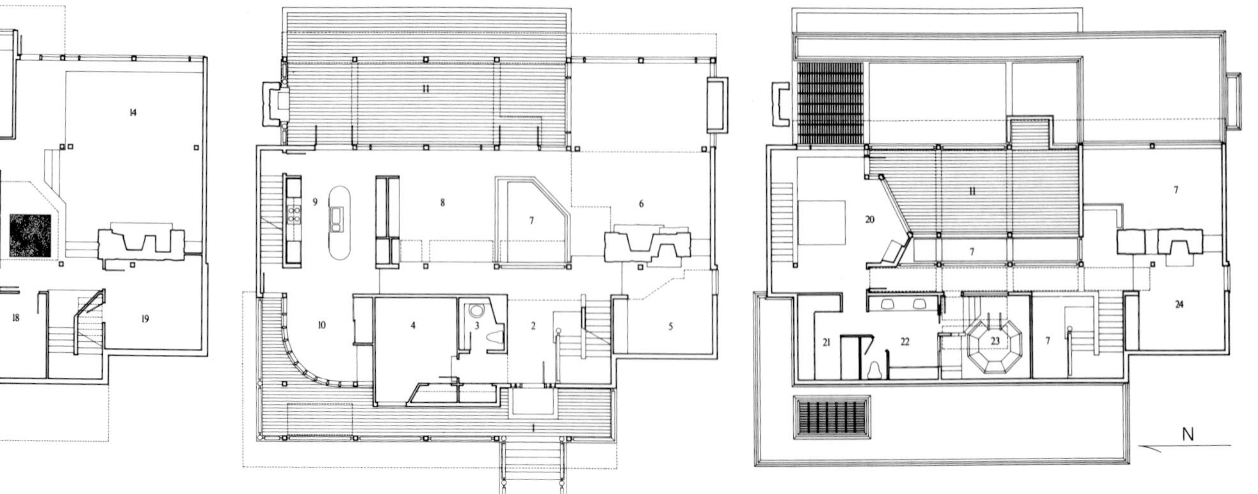

Northern Prairie House

Formal order and careful detailing in a large home for a family of four

by John Patkau

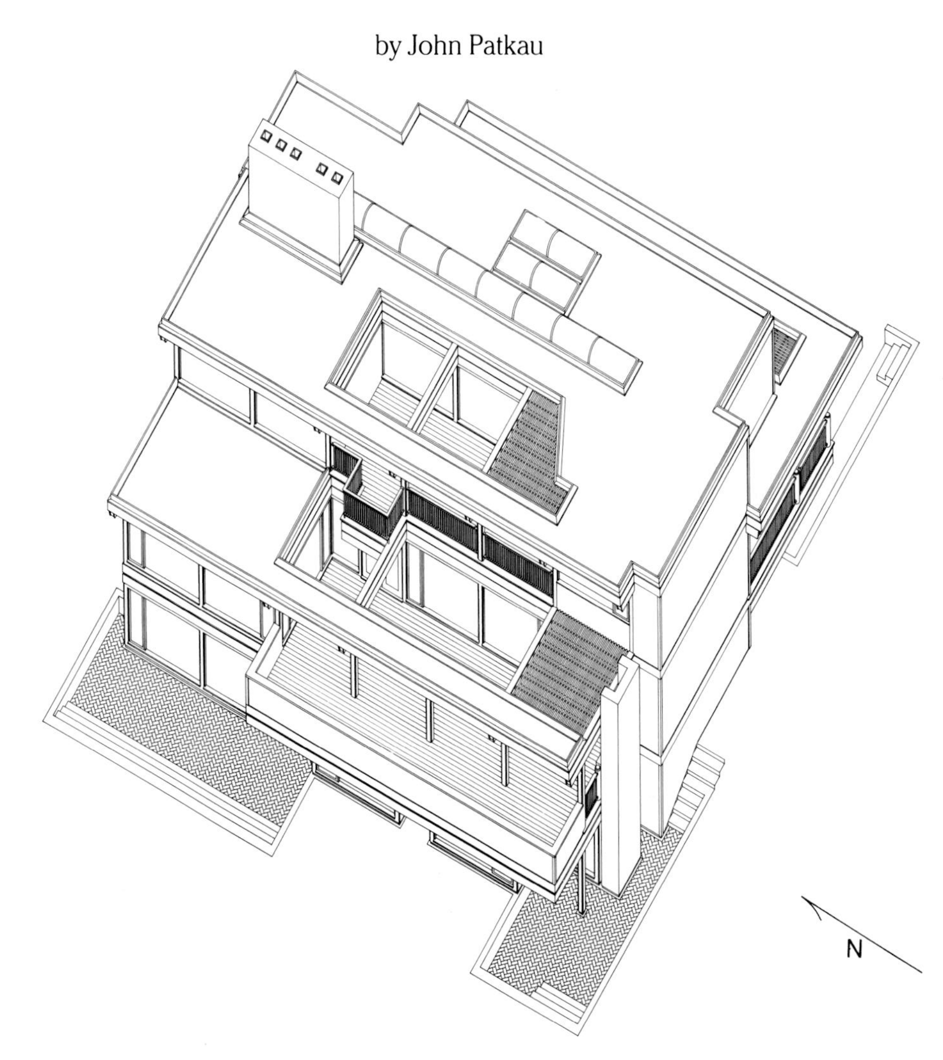

Edmonton is the most northerly major city in Canada. It is in the province of Alberta, at the very top of the great plain that sweeps down through the heart of North America. Historically, it has been the capital of a rich agricultural region, but during the past 30 years it has witnessed a booming petroleum industry.

Stu McGregor grew up in Edmonton, and purchased a beautiful lot on the southwestern edge of the city. Wooded, with a light screen of poplar trees and wild rose bushes, the lot overlooks the broad valley of the North Saskatchewan River. It slopes gently from the street to the edge of the valley and then drops precipitously down to the river below. Across the river the valley is a panorama of pastoral tranquility.

Stu and I first met in the summer of 1979. He had approached me, and the architectural firm in which I am a partner with my wife Patricia, to discuss the possibility of designing a house for his family. Stu and Barb McGregor have two children, and they wanted a large house that would include a variety of areas for family recreation and entertaining. These areas were to be different in size and spatial quality to accommodate the requirements of different types of activities. Our goal was to design a house that would sustain a range of current family activities, as well as the changing and developing activities of the family as it matured.

Siting and organization—Because of the instability of the steep bank of the river valley, the actual area of the lot on which a house could be built was small. For this reason, we arranged the house on three levels: the main level is dedicated to family activities, the lower level to the children, and the upper level to the parents. The house is entered from the street side at the main level through a broad covered porch (photo above left). Beyond the porch, the house is closed to the street for privacy. On entering the house, you immediately see the river valley beyond, for the opposite side of the house is almost entirely glass.

To order the interior spaces of the house, each of the three levels is organized about an axis that runs roughly north/south, from one side of the house to the other. This axis takes the form of a hallway that connects the various spaces on each level. The design is reminiscent of architect Richard Meier's work. Meier frequently arranges his homes on this kind of formal, axial basis.

Along the north/south axis in the McGregor house, a 10-ft. structural module regulates the spaces and the rhythm of movement within them (drawings, facing page). At each end of the hallways, a stairway perpendicular to the hall connects the three levels. An irregular struc-

James A. Dow

Horizontal continuity is the primary function of the trim in the McGregor house. In this view down the hallway on the third level (above), the various elements start at floor level with the baseboard, followed by the 5-in. wide header molding atop windows and doors. Above this, 1x12s separated by a ½-in. gap reveal the placement of the structural members. On the second level (top), the boxed-in posts and beams define the outlines of various spaces within the house. The sitting areas in the foreground are governed by the placement of the beams, as is the dining area in the background.

tural module in the east/west direction allows the rooms to assume the dimensions appropriate to their functions.

Materials and scale—In a climate where winter lasts five months or more, building materials that look warm and rich have a natural attraction. In Edmonton the most readily available materials with these characteristics are western red cedar and a broken-faced siltstone called rundlerock. The McGregors wanted to make extensive use of both of these materials. This posed something of a problem for Patricia and me because we believe that these materials are frequently used improperly in house construction. Often they are applied in a superficial manner, like wallpaper, simply for effect.

This was our first house commission and the first time we had used cedar and rundlerock, so to prepare ourselves we studied buildings in which similar materials had been used in architecturally powerful ways. In particular, we studied the California bungalows of Greene and Greene and the Ise shrine in Kyoto, Japan.

In the Greene and Greene houses, wood is used in a sensuous manner that highlights its color, texture and grain. The Greenes used wood details that are related to one another in shape and scale, and repeated them throughout a home to create pleasant, coherent spaces with a tangible sense of integrity.

In the Ise shrine, wood is used monumentally. This ancient shrine, which has been rebuilt every 20 years since its initial construction, is a post-and-beam structure proportioned to create a bold and immediate impact. Columns, beams and even newel posts are larger than they have to be for structural purposes, which gives them a presence that is hard to ignore. At the same time, every member is rendered with such preci-

From *Fine Homebuilding* magazine (April 1986) 32:24-28

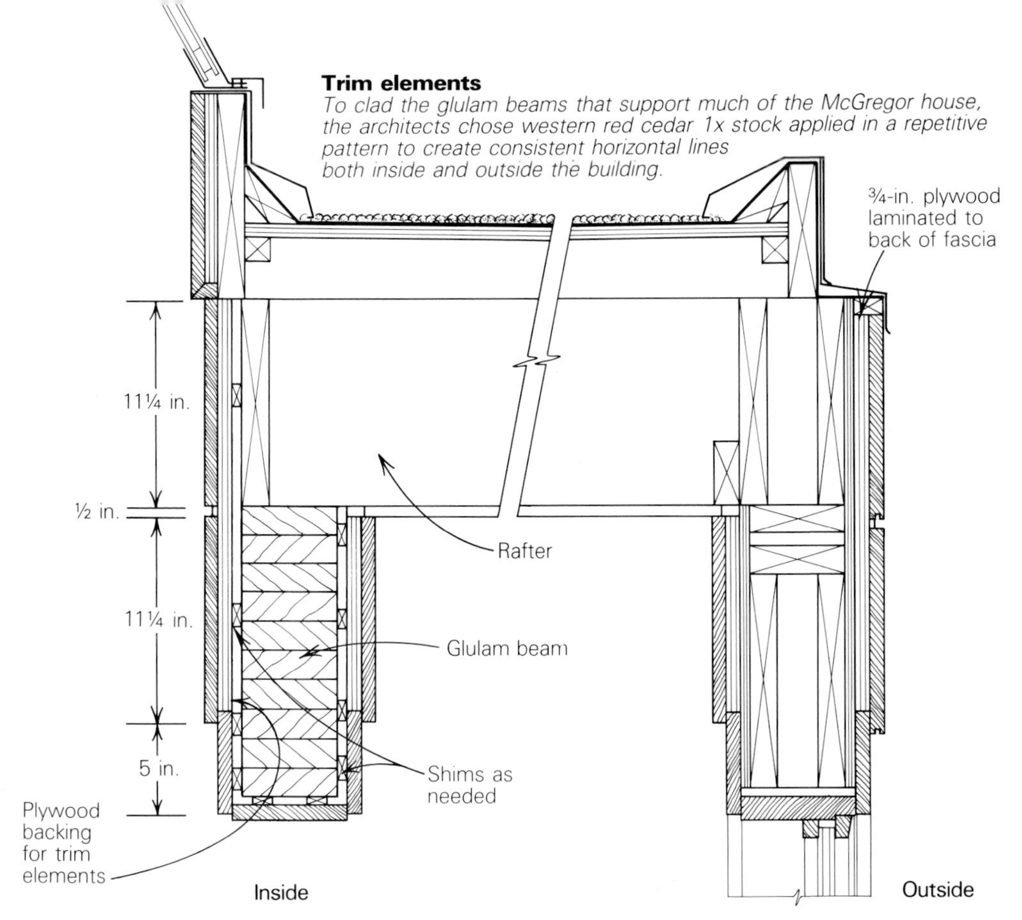

Trim elements

To clad the glulam beams that support much of the McGregor house, the architects chose western red cedar 1x stock applied in a repetitive pattern to create consistent horizontal lines both inside and outside the building.

Cylindrical posts at the stairway and guardrail contrast with the predominantly rectilinear components of the house. Rounded handrails are attached at an angle that makes them easy to grasp. Carborundum strips improve footing in icy weather and lengthen tread life.

sion and refinement that a sense of timelessness pervades the entire complex of buildings. Together, we felt, qualities such as these would provide us with a complete building vocabulary.

Systematic detailing—To achieve some of the qualities we admired in the buildings we had studied, we developed a system of detailing that would assemble such diverse elements as window and door casings, beam and column cladding, baseboards and other miscellaneous trim elements within an overall discipline. The system is based on the consistent use of five horizontal elements.

The first of these elements is a 5-in. wide band that runs at a constant elevation of 80 in. on each level, tying together all door and window heads. Immediately above this, 11¼-in. wide boards clad all beams framed below ceilings. Beams in the north/south direction are given dominance over beams in the east/west direction to reinforce the axes about which the house is organized. In areas where ceilings are dropped, they are dropped to the top of the first element. In areas where ceilings are raised, the first element is retained as a picture rail on the wall to maintain continuity and a sense of scale (photos facing page). Where openings are created through floors or roofs, and where the edges of floors or roofs are expressed, two further elements are added: another 11¼-in. band representing the framing of the floor or roof assembly above the beam (drawing, above), and one of a variety of bands above it to represent the framing or parapet. The final horizontal trim piece is a 5-in. baseboard. Except for the baseboards, these horizontal lines prevail outside the house as well.

Vertical elements such as column cladding, door and window casings and corner boards are then simply framed between these bands of trim. This allows for butt joints at every intersection, which is especially important outside, where seasonal humidity fluctuations cause the wood to expand and contract. This action will work away at mitered joints, eventually creating awkward gaps between trim pieces.

This system is very straightforward in theory. In practice it demands precision and impeccable workmanship. Fortunately for us, the contractor, Ed Lusis, is a craftsman. Working with beautiful clear cedar of a rich honey color, Lusis executed our intentions better than we could reasonably have expected. On a number of occasions he was able to suggest technical improvements to our details. For example, the 11¼-in. wide bands that are the major elements of our detailing were originally intended to be 1½ in. thick. Lusis was concerned that even though they were kiln dried, their hefty size and the dryness of the air in Edmonton would cause them to cup once they were in place. To avoid this, he nailed and glued ¾-in. thick material to plywood backing. These trim elements were then attached to the structure using shims to compensate for irregularities in the thickness of the framing members. After four years these pieces of trim have remained perfectly flat.

Another of Lusis' touches turns up along the underside of each exterior fascia and band board. On each of these he cut a shallow kerf to help keep rainwater from wicking behind siding, soffits and porch ceilings.

In addition to the large cedar trim pieces, the entire exterior of the house is clad in 1x4 cedar siding applied vertically. Most of the ceilings are also of this material. All of the cedar was prestained with a homemade mixture known as "Vic's formula." Victor Fast is an Edmonton architect, and he uses a mixture of 98% Pratt and Lambert clear exterior finish blended with 2% pigmented stain in a butternut color. This concoction adds intensity to the color of the wood without overpowering it, and allows the cedar on the exterior to take on a subtle silver-beige hue as it ages.

Because cedar is very soft, we used clear fir where we expected wear or needed greater strength. Door and window frames, wood decks, exterior stair treads, handrails and newel posts are all fir.

Newels, rails and treads—The handrails and especially the newel posts are influenced by our study of the Ise shrine. To what would otherwise be an essentially planar design these overscaled elements provide contrast, and their rounded shapes are more inviting to the touch. Like the newel posts, the 2x6 handrail is rounded to fit the hand. The handrails both inside and out are affixed to the baluster rails at an angle (photo above). This tilt allows the hand to get a better grip on the railing than that allowed by most squared-off handrails.

The fir stair treads outside the house were a special problem, because they become very slippery in the frigid Edmonton winters. Add a light layer of snow or ice and they become treacherous. To remedy this problem, we inserted a double row of abrasive Carborundum strips in grooves near the front edge of each tread. Also, these strips substantially reduce the wear caused by foot traffic, weather and snow removal.

Fireplaces—We used two rundlerock chimney masses to anchor the flowing spaces of the house to the site and to provide a counterpoint to the regular order of the post-and-beam structure. The first rises within the house through all three levels, allowing fireplaces in the recreation

Several fireplaces made of a siltstone called rundlerock contrast with the cedar trim and ceilings. Each fireplace, such as this one on the second-level family room, has a hearth and mantel made of a single stone.

room, living room, family room and den. The second contains a barbecue, and is located to the side of the large main-floor deck; it visually balances the central-fireplace mass.

The weighty presence of masonry provides the sense of permanence we wanted. At the same time, the broken-faced ruggedness of the rundlerock gives the house a contrasting surface that enhances the silky richness of the cedar.

To reinforce this ruggedness and mass, we looked for large rocks for mantels and hearths (photo above). We had no trouble finding them, but we had to bring a crane onto the site to lift them. This proved to be quite a spectacle, as the whole crew was needed to guide these mammoths through chimney openings in the framing and lever them into place.

Flat roof—Building in a severe northern climate does not require techniques other than those used generally throughout North America. It does, however, require great care in the design and construction of the building envelope. Mistakes can lead to costly problems like frozen or burst water pipes, or at the very least a loss of comfort and high energy costs.

We chose a flat roof for the McGregor house because the building's details—concentrated, as they are, on the walls—emphasize its horizontal nature. A pitched roof would have negated the effect we were striving for. Fortunately, Edmonton's climate is on the arid side, with snow loads that are light enough to allow a flat roof if it is done correctly.

Typical deck vent

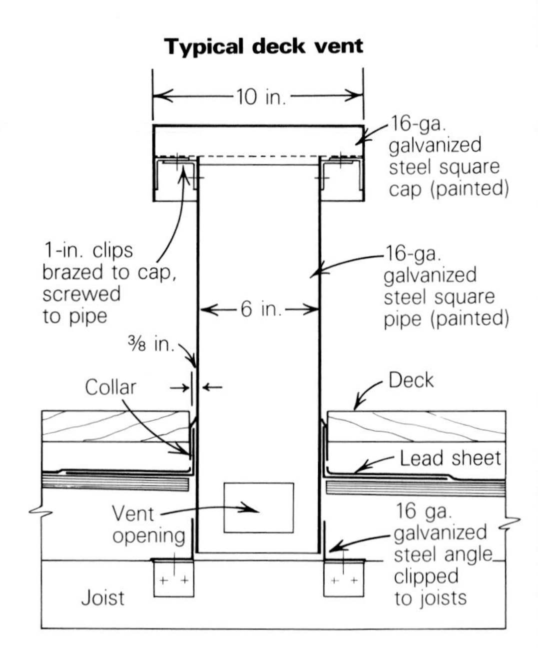

The most important thing we did was to provide positive drainage by placing tapered sleepers directly atop the 2x12 roof joists. They slope toward the center of each roof, where a drain picks up rainwater or melted snow and conveys it to the storm sewer via pipes that pass through the interior walls. This prevents frozen runoff from accumulating in exterior drains.

Because we were using 12 in. of fiberglass-batt insulation between these joists, the sleepers also made for a continuous vent space above the joists so that any moisture-laden air within the structure could escape before the moisture in it condensed. Proper air flow through this vent space is critical in cold climates. In this house, we were able to get the required air flow with a combination of soffit vents, in conjunction with specially made mushroom-shaped roof and deck vents (drawing, above).

The roof is a four-ply hot-mopped built-up membrane topped with gravel. It's applied over ¾-in. T&G plywood decking. This house has two roof decks over living spaces. To level the decks, the 1x decking planks were nailed to sleepers that taper in the opposite direction from those that create the roof's drainage slope. Unlike the rest of the roof, the membranes under the decks are covered with a heavy layer of felt. The deck sleepers bear on this layer of felt.

The final component of the roof or roof-deck assembly is a continuous vapor barrier, located on the underside of the joists, in which all joints have been sealed to prevent moisture from within the house entering the roof cavities. Any electrical fixtures that are recessed into this assembly were carefully boxed off with adequate air space provided to allow heat from the fixture to dissipate. Continuity of the vapor barrier was maintained around the recess. □

John Patkau is a partner in Patkau Associates/ Architecture and Interior Design in Vancouver.

The Art of Workmanship

Designing with a modest budget and basic construction skills

by Llewellyn Seibold

There's an adage that states that the best things in life are free. My wife and I have found this to be true, particularly in houses we admire. Such houses have an abundance of light, fresh air and simple provisions for family comfort. After living in many houses that lacked these basic ingredients, we decided to build a house for ourselves that would include them. Our goal seemed so basic and the success of such construction so illusive that we suspected hidden problems must lie ahead.

As an architect, I had some understanding of the complexities involved in making buildings. At the same time, I was dismayed by the observation that architects often seem more like managers or administrators than designers. The modern architect, unlike such architects/craftsmen as the Greene brothers, Bernard Maybeck and Henry Mercer, often loses touch with the actual work going on. Perhaps, I thought, the key to success with this project was to design *and* build the house. This approach eventually led to a profound change in my thinking about design and construction.

A link to reality—The simple vision we had for our house quickly met complications as we started to consider how it would be built. Our appreciation—and expectation—of careful craftsmanship and thoughtful detail seemed unrealistic as we considered our modest budget and limited skills. It was clear that our resources would be stretched to the limit.

Along with my design background, I had enough building experience to know that "simple" structures are not always simple to construct. Considering our budget, we decided to build as much of the house as possible using our own labor, which led to some unfamiliar ways of thinking about the design. Given our very basic construction skills and the limited tools that we possessed, I had to think in terms of realistically achievable workmanship, not perfection.

About this time, I discovered a book by David Pye called *The Nature and Art of Workmanship* (Cambridge University Press, 32 East 57th St. New York, N.Y. 10022, 1968. $15.95, paperback). The ideas Pye presented proved to be the missing links between our dreams and the reality of building a house. He offered a new way of thinking about the quality of materials and products, and it was reassuring to realize that quality work is not dependent on exquisite materials or highly skilled workmanship. For the first time I began to understand that good workmanship was not exclusively the result of mystic knowledge, a limitless budget or an unrestrained schedule. This was the real starting point in the design of the house.

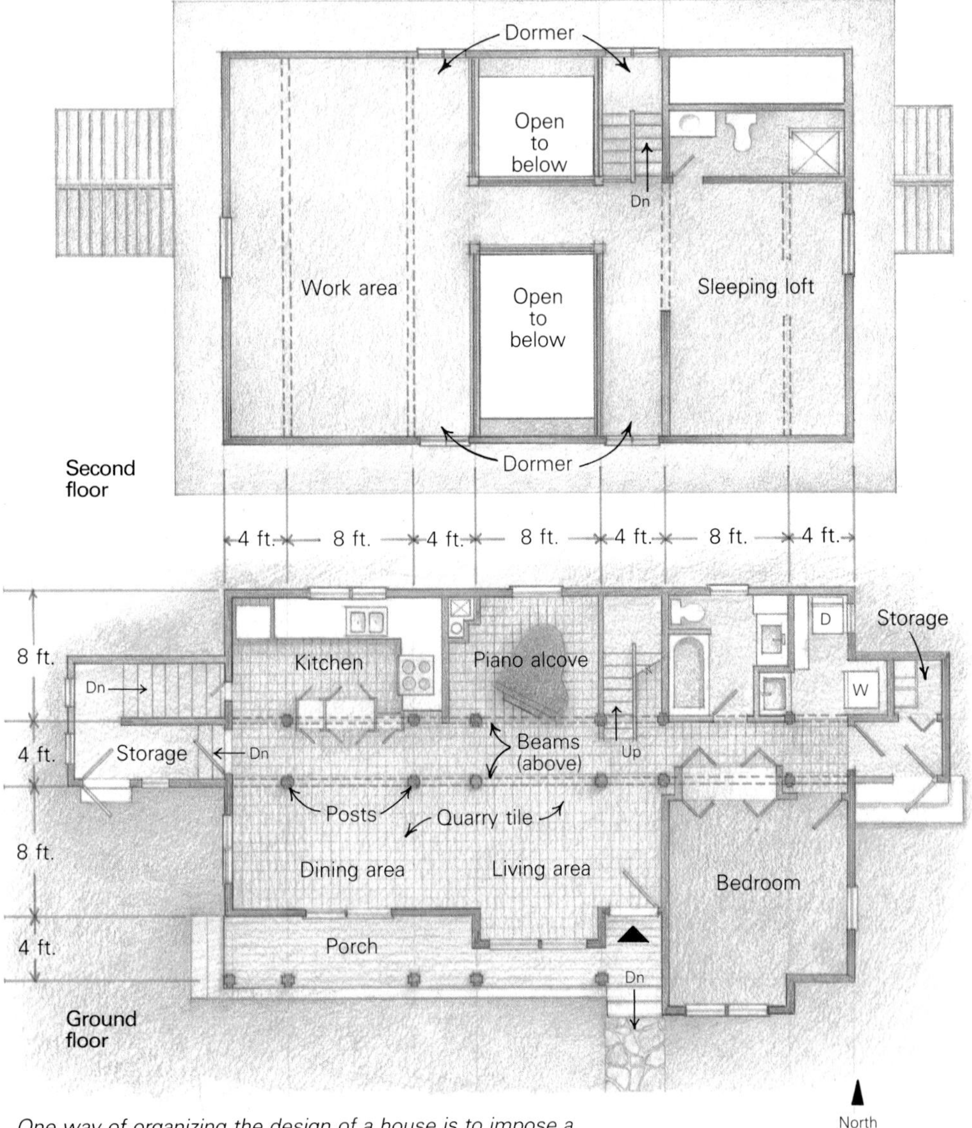

One way of organizing the design of a house is to impose a uniform system of measurement that dictates the location of various elements. In this case, 4-ft. and 8-ft. modules determined the location of elements such as posts, beams, windows and doors.

Workmanship, certainty and risk—Pye noted two concepts which are central to understanding workmanship. One is "workmanship of certainty," the other "workmanship of risk." The workmanship of certainty is common in industry today and involves the use of jigs, templates, and other shape-determining devices. Results are predetermined if the tool is set up properly. Most tradespeople tend to move toward this way of working, which is understandable because it diminishes the chance for error and is faster, particularly when the work is repetitious. The workmanship of risk, on the other hand, involves the potential for uncertain outcomes. The quality of the result is continually at risk, and judgment, dexterity, and care must be exercised as the work is performed. Throughout the project, a real possibility of spoiling the work follows the worker like a shadow.

Well-built homes today aren't necessarily made solely by one type of workmanship or the other. The work usually employs both processes. In many cases a worker will move back and forth between "risk workmanship" and "certain workmanship" without much deliberation or delay. Materials produced under highly regulated factory conditions are assembled in the field under less than ideal circumstances, with a lot demanded of a worker's judgment during the installation. Even with the increasing shift toward systematized, modular construction, weather conditions and site irregularities are unpredictable. In our project, I began to see that the house was a hybrid of hand labor and industrially produced components. In fact, the real design and budget challenge were to orchestrate the interaction of these different ways of doing the work in order to maintain an overall design coherence. Risk workmanship produces things with diversity and variation, while certain workmanship produces things of exactness and predictability. Both of these elements are important and create in good design a hybrid vigor stronger than the simple sum of attributes.

Refining the design—We decided on a simple volume that was as high at the ridge as it was wide, with a length not quite twice as long as the width (24 by 24 by 40). This proportion came in part from the sum total of needs to be provided for and in part from the wonderful proportions of Shaker buildings. At first we imagined that the house would be like a cottage as soon as we completed it. We visualized a modest, careful house set in a garden laced with paths and surrounded by

The simple structure of the house focuses atteneion on the details. Slight reveals created by built-up rafters and purlins (above) create shadow lines that accent the broad planes of the roof, and built-up beams echo the reveal (photo, p. 47). Dormers on the house are unusually large and provide niches for additional work space. The railing is simply painted steel pipe; the black escutcheons are rubber washers intended for toilet tanks. Flooring upstairs is oiled Masonite.

In keeping with the theme set by the rest of the house, the exterior is spare yet carefully considered. This creates an overall feeling that is solid and forthright, yet contains a degree of fineness and lightness. In the photo above, purlins cantilever to support the eaves in a detail that reduces the complexity of the roof structure. In keeping with the modest budget, exterior materials are modest and straightforward: shiplap-pattern Masonite siding, metal roofing and vinyl-clad wood windows.

an orchard. The garden and the orchard have been started, but we've learned to temper what we see in our minds' eye with a healthy dash of reality. After all, a landscape is a thing requiring even more patient workmanship than a building.

The house was proportioned and dimensioned on alternating 4-ft. and 8-ft. modules. The rhythm of the modules regulates the placement of columns, beams windows and dormers. The ground floor contains the major living spaces, while the upper floor is devoted to sleeping and work spaces for myself and my wife (see floor plans). I added two appendages to each end of the house to accommodate an entrance to the carport and garden, access to the basement and some storage. The house has a partial basement that doubles as storage and a refuge from Kansas tornadoes.

Inspiration for the design came from numerous sources. I had come to appreciate the straightforward and well-built vernacular buildings of central Kansas and was especially influenced by the direct expression of structure and utility in the great barns of the northern prairie—I am indebted to them for a good deal of my structural and aesthetic education. The simplicity and order of Shaker settlements, which exemplified high ideals in early American building, appealed to me as well. The Arts and Crafts tradition, Japanese folk houses and Scandinavian design were further sources of inspiration. I have come to use these sources as standards against which to measure the results of my own work (and the work of my students) in the design of small-scale buildings such as houses.

A structure of simplicity—The framing of the house was based on the logical sequence of members from columns to beams to joists to rafters to purlins, which is a pattern found in many barns as well as in Japanese folk houses. This principle of layered construction was adapted to use commonly available dimensioned lumber, assembled by a couple of people with basic construction skills. While the feel of the structure inside and out reflected the essence of agricultural buildings, I wanted to make sure that the house would feel more like a house than a barn. We decided to scale the construction materials to the size of the project.

Each column was composed of five 2x4s nailed together to form an interesting shape in cross section. The major beams consisted of two 2x12s nailed together like a standard header. Each roof beam was a pair of 2x10s spaced apart 1½ in. to accommodate floor joists and collar ties. The purlins were pairs of 2x4s nailed together with a 12-in. plywood spacer between them. We left the spacing exposed between members, which resulted in a play of alternating light and dark lines throughout the structure (photo above left). This articulation resulted in an overall feel which is solid and forthright, like barn structures, but also has a degree of fineness and lightness that the Japanese folk houses possess. Other materials were equally straightforward, including the shiplap-pattern Masonite siding and metal roofing (photo above right).

The cold sweat of reality—To organize the construction and prepare an estimate of building costs, I completed a very limited set of working drawings. The drawings included floor plans, elevations, a building section and some structural details. I left many details to be worked out on site. In most cases, this led to better solutions than I could have anticipated, though it did lead to some headaches during construction. Looking back, I should have given more attention to the mechanical and electrical design.

For the first few months, I had hired a young carpenter, Gail Martinson, to help us. Although the house was different than any he had worked on before, his deliberate procedures and insights were invaluable. After construction started early one summer, the schedule was very tight because I had to begin teaching late in August, but progress was slower than we anticipated because we ran into delays in obtaining materials. Our roofing supplier had guaranteed the material within 10 working days from the time it was ordered. The metal and flashing arrived at the end of that period—without the fasteners.

During this time, Pye's philosophy about materials and workmanship became increasingly important. He described the use of English walnut, which is worthless firewood until it is sorted, graded, sawn and finished. The raw material may promise certain things but it is workmanship which gives it quality. This idea was liberating. I recall the dismal moment when we cut the strapping around a bundle of 2x4 cedar; boards jumped in every direction because they were warped and twisted. They might have served as firewood come winter. But by combining individual pieces into composites, and by taking good care in selection, greater refinement and reg-

Any material used to excess can detract from a design, so the kitchen features European-style laminate cabinetry as a contrast to all the wood in the house. Wood edging on the cabinets and countertop maintain a visual tie to the rest of the house.

Most floor surfaces downstairs are quarry tile. Support posts are built-up from 2x4 stock (photo at left). The stair details illustrate the premise that a fancy thing done badly is not craftsmanship at all, yet a simple thing done well is the essence of such work. Mitered nosings on the stair would have called for more skill than was available, so the connections were butt-jointed instead (photo above). Maple decorates the joint, and button plugs (which are easier to install than flush plugs) conceal countersunk screws.

ularity was given to this material: the jumpy 2x4s became columns which were easily put together and interestingly shaped. The 2x4s were simply nailed together with 12d galvanized nails. At first I thought this might appear to be too crude, but the regularity of the nailing pattern, the dull finish of the nails and the effort we went to in setting each nail made an orderly and pleasing composition.

The workmanship of balance—My biggest concern from the start was how the columns and beams would appear against the flat, smooth surfaces of the drywall. One material (wood) would shrink, swell and have less then perfect features. The other material (drywall) would be dimensionally predictable, with square edges and even features. In the end, the joints between them were acceptable where they showed the most, but in some places they could have been better.

We decided to balance the pervasive use of wood in the house by using plastic-laminate cabinetry. Wood edging around the countertops ties the laminate visually to the rest of the house (photo top left). Because of the regular and precise nature of laminate, it demands a workmanship of certainty. We decided to have the cabinets built in a shop which had the tools and skills necessary to do the work in accordance with the demands of the material. It is important to choose carefully a design that will unify materials and an appropriate workmanship.

I feel that what makes building different from other crafts, such as cabinetmaking, is that in building, one must be able to move with ease between the workmanship of risk and the workmanship of certainty. Building is a complex orchestration of budgets, site circumstances, schedules, codes, skills and the tendencies for humans to err, and doing what is appropriate is a tremendous task. We learned this as we were putting on the metal roofing. We had earlier squared and temporarily braced the roof beams, but in our zeal to get the decking on, we failed to double-check everything. Too late we discovered that the wind storm that blew across the site one night had pushed the roof out of square. If we had used shingles or tile roofing, this would have been a negligible problem. With metal roofing, however, there's little tolerance for error because the material is produced under a workmanship of certainty and requires precise installation. And on a simple roof like this one, problems of detail were exaggerated. This was a forceful reminder of the idea that simple forms are not necessarily simple to build well and showed us how a workmanship of certainty—the roofing—was dependent on a workmanship of greater risk—the framing. We learned the hard way that much of building and design is knowing what to fuss about and what is a waste of time and energy.

The drywall finishing, too, was compromised during construction because of incompatible levels of workmanship. Drywall is, of

Photos this page: Llewellyn Seibold

Recognizing a restricted budget and limited construction skills, the owner/architect planned the house around simple materials and basic joinery.

course, a material of regularity and flatness, but the joints aren't so. The drywall crew was working on the house at night during off-hours. I had electricity in the house, but most of the lights were not usable, so the finishing turned out to be rough in places where it should have been precise.

Principles to apply—Upon some reflection, I've come to understand several principles that guide my consideration of a building's appearance, the material selection, and the process of putting the pieces together. These principles help me to balance every vision with the dust and sweat of realizing it.

First, I attempt to determine early on what parts of the design and construction demand focused attention and care. This probably becomes second-nature with most builders, but it deserves mention because it may call for a very uneven distribution of budget and skills during the course of a project. For example, a quarry-tile floor covers most of the major spaces in this house, and the tile we used was very consistent in color and size (photo facing page, bottom left). The grout work, however, is of inconsistent quality and this compromises the effect of the whole. Knowing in advance that our grout work would lack the consistency evident in the tile, we could have installed thick, irregular tile of varied coloration so that the grout work could be much more free in its workmanship and be acceptable. In staying with the same tile, it would have been better to be less compulsive about laying it out in order to spare effort for achieving greater regularity in the joint work.

Second, I use segmented joinery where skill levels are uncertain or minimal. This leads to the appearance of higher quality. The effect is similar to the methods that turn-of-the-century carpenters used when casing a door or window. Intermediate pieces of material were interjected between the principal pieces to change direction or width. For example, we used a small bead between the side casing and top casing of doorways, and plinth blocks at the window casings. The cherry nosing on the stairway is not mitered but is broken by a small bead of maple and rounded as it turns the corner (photo facing page, bottom right). Square cuts are easier to do, and this became a way of dealing with my lack of building skills and sophisticated tools.

Another form of segmented joinery is the use of trim to mask the differences between materials. The joint between ceiling drywall and a principal beam was fitted with ½-in. quarter round, which can accommodate and conceal the irregularity between both surfaces.

Finally, I balance simple skills and materials with those that are more complex and ornate. Rarely does one encounter unlimited resources in money or skills. I sometimes wish that everything I design or build could be crafted like a piece of fine furniture, but I also have an egalitarian sense which tells me that craftsmanship need not be the exclusive privilege of those who have power. There's a vitality that comes from using highly regulated and common surfaces, such as drywall, in combination with freely worked materials like wood. In this house, the combination of plain walls and different colored woods enriches both the plainness of the wall and the coloration and pattern of the various woods (we used pine, fir, cedar, cherry and maple).

It's possible to obscure excellent workmanship by overusing a particular material. When expensive materials and labor have been used, this becomes very disheartening. I find it best to let one material or system dominate another in order to establish a visual hierarchy. The stairway in our house is a pleasing and regulated construction of maple, fir, cherry and birch, but it is set in a simple surround of wood framing and white walls. The red quarry tile on the floor is related in color range to the wood, but it contrasts well with the plain wall surfaces. The whole atmosphere of the house takes on a warm glow that is not overpowering because it is balanced (photo above). It makes the place memorable and cheery, and filled with light, air and the hearty sense of home. □

Llewellyn Seibold is an architect and a teacher of architecture at Kansas State University in Manhattan, Kansas.

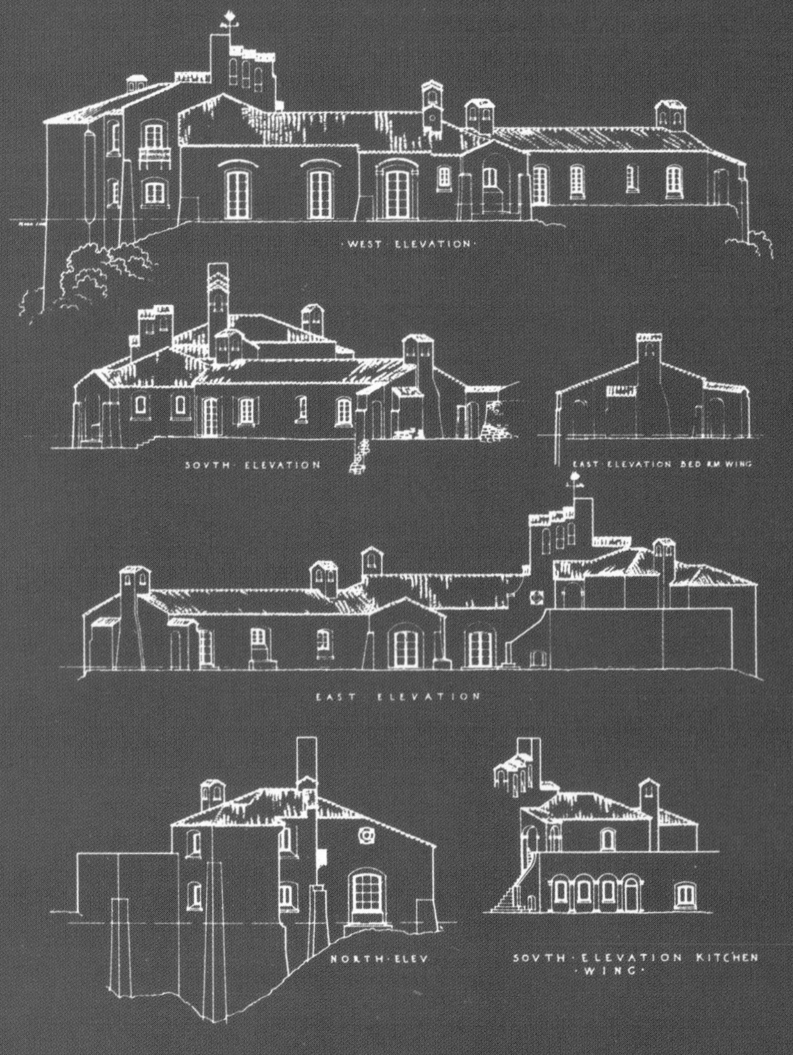

·WEST ELEVATION·
SOUTH· ELEVATION
EAST·ELEVATION BED RM WING
EAST ELEVATION
NORTH·ELEV
SOUTH·ELEVATION KITCHEN ·WING·

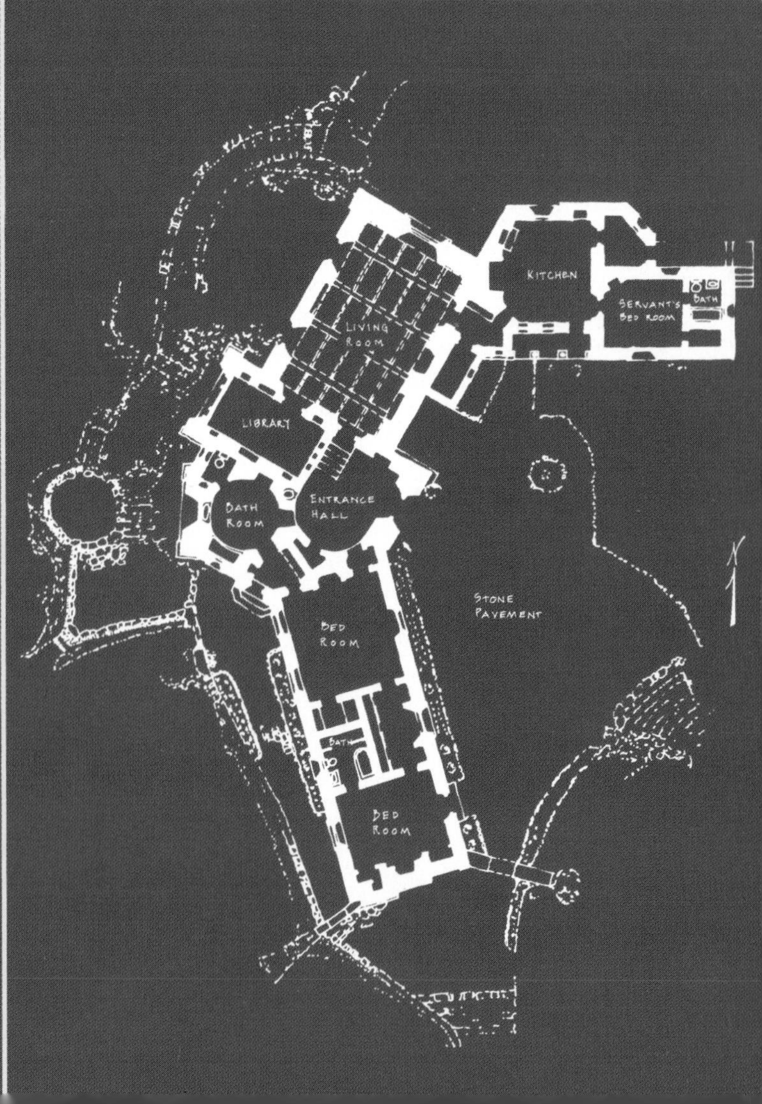

KITCHEN
SERVANT'S BED ROOM
BATH
LIVING ROOM
LIBRARY
BATH ROOM
ENTRANCE HALL
STONE PAVEMENT
BED ROOM
BATH
BED ROOM

The James House

Charles Greene's masterpiece in stone

by Charles Miller

Architecture's most famous sibling partnership came to an end in 1916, when Charles Greene moved to Carmel, Calif. After designing homes in Southern California for three decades, Greene left Pasadena for a quieter life on the seacoast to the north. Although he retired from the partnership with his brother (see p. 58), he continued his architectural practice on a smaller scale, receiving occasional commissions for houses and remodelings. One such commission stands out as the triumph of his later years—the James house.

D. L. James was a well-to-do merchant of fine china and glass from Kansas City, Mo. He was a lean, athletic man of great personal charm and elegance, and a graduate of Yale and Oxford. The china business provided a comfortable living, but it wasn't conducive to James' aspirations—he had a burning desire to write for the theater. He set summers aside for writing, and he spent them with his wife and son on the West Coast, near Carmel.

In the summer of 1917, James purchased one of the most spectacular and rugged building sites on the California coast. This rocky promontory is a granite triangle, split by crevices into fingers that reach 80 ft. to the tide pools below. The sea bottom is mottled with dark, eroded boulders and brilliant quartz sandbars. The sand forms a bright backdrop, which rebounds the sunlight through the clear water, creating an array of blues from turquoise to indigo. Rust-colored kelp beds twist in the swells, and the cliffs absorb the waves' impact with distant rumblings. It seemed the perfect site for a writer's retreat.

On their next trip West, the Jameses went to a tea party hosted by some friends. Among the guests was Charles Greene. As the afternoon progressed, James and Greene became acquainted, and talked about the granite site. The next day, Greene visited the site and did a rough drawing of a house. He showed it to James, who liked the idea and promptly retained the architect to design a summer house. Thus began, with little ceremony and no preconceptions, one of the most magnificent houses in America.

Influences—By the time James met Greene, the architect's successful Pasadena years with his brother Henry were behind him, and residential commissions were hard to find. People in the market for a custom home wanted Spanish Colonial Revival, French Provincial or Tudor—almost anything but the Arts and Crafts-style wide-eaved wooden houses made famous by the brothers. But James was unencumbered by any stylistic baggage. In fact, he'd never heard of Charles Greene. He liked and trusted the man, and that was enough.

Greene began his planning with a contour map of the site. Its natural features suggested a building with a footprint shaped somewhat like a truncated V. This form wrapped around the point, taking advantage of the only piece of the lot that could be described as level. Greene placed the living room on this flat, where it joins the two wings of the house (drawing, facing page, right). They roughly follow the contours to the north and east.

Like other building sites on the open coast, this one is subject to ferocious weather. Winter storms sometimes reach hurricane force. But the typical weather pattern is a calm morning, followed by prevailing afternoon breezes off the ocean. Wrapping the house around the point creates a courtyard sheltered from the wind.

Greene must have wrung his hands with joy when he saw the local building supplies. His Arts and Crafts background encouraged the use of natural building materials, and here he was surrounded by sparkling golden-hued granite on a site that demanded a weighty presence. The brothers had used stonework

The site and the plan. **Anchored by battered walls tied to bedrock, the James house rises like a medieval castle perched above its fiefdom. A 1923 view from the south, above, shows the stone aprons that extend down the cliff. Greene roughly followed the contours of the site as he laid out the wings of the house (drawing facing page, right). He placed the living room at their intersection, where it opens onto a terrace 80 ft. above the ocean. From the northwest (top), the roofs step down from the massive central chimney. Greene's presentation drawings (facing page, left) reveal an even more ambitious plan. Here, stairs rise from the courtyard to a second-story study, flanked by a terrace over the kitchen. This section was never built.**

From *Fine Homebuilding* magazine (December 1984) 24:26-32

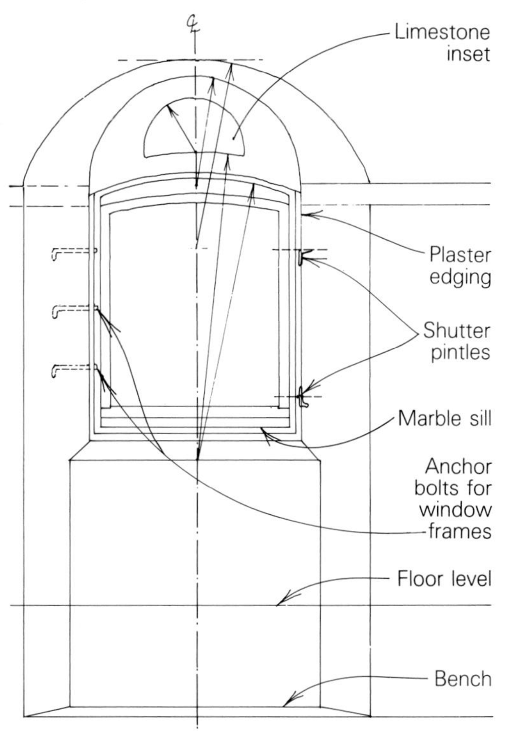

Bathroom gable
Greene combined thin, horizontal rocks with larger cobbles in the bathroom gable (facing page). He elaborated on this form with built-up corners, deep reveals around the window and an arch that springs from various centerpoints as it moves from the top of the window frame to the plane of the exterior wall (above).

in many of their earlier homes to accent other materials. They were especially adept at blending rounded river rock with the tortured shapes of clinker bricks. But they had never attempted a house built completely of stone. Greene's presentation drawings of the elevations show little of the rocky textures that would distinguish the James house, but the proportions are there, carefully drawn and waiting for the masons (drawing, p. 52, left). The eaves are clipped, presenting little resistance to high winds and admitting as much light as possible. Each window and door is capped with an arch, echoing the eroded natural arches that support much of the site. The house rises in a series of terraced tile roofs, peaking at James' second-story study.

Work began in 1918, with Greene personally supervising construction. He'd taken up residence in Carmel, so a daily trip to the site was easy for him. He found a mason named Fred Coleman, a hard-working man who could take directions, and he hired some laborers to help bust rock and carry hod. They found a quarry site a mile down the coast and began hauling horse-drawn wagon loads of fractured granite to their precarious site.

From bedrock to building block—Nowhere is there a sign of a manmade foundation under the James house. The walls splay out to meet the bedrock, and the juncture is indistinct (photo bottom right). Coleman and his men hollowed out a groove in the granite, and began the exterior walls with large rocks. These first courses are made of weathered surface rock, chosen by Greene to ease the transition from the bedrock to the walls. The mortar was held back from the faces of these first courses, and the resulting gaps were filled with dirt—an earthen grout line.

In places, the walls are over 4 ft. thick. Their insides are a jumble of rocks and mortar, and they are topped by double 2x12 plates. The plates are anchored by ½-in. rods that extend 4 ft. into the stonework, where they are bolted to 4-in. by 5-in. steel plates.

It is the texture of the exterior stonework that puts the James house into a league of its own among stone buildings, and its assembly was an exercise in tenacity. Coleman would grip a chunk of raw granite in his bare left hand. Then he would cleave it into finished pieces with his mason's hammer, and Greene would tell him where to put the pieces. Each rock is arranged with its thin edge to the weather. The mortar joints are deeply tuck-pointed. In some places the stones are only ½ in. thick, while the mortar lines around them are triple that. From a distance, the effect is like a swarm of thick brush strokes aligned by magnetic forces.

Greene kept constant watch over Coleman, to make sure the walls had the right texture and balance. Just when the surface was about to become repetitive, he would have Coleman start a course with thick, cobble-like stones that define a new horizontal line, and help to prevent vertical cracks. Once Greene had to go to San Francisco for a few days, and returned to find a freshly built wall section that didn't meet his approval. Rather than live with it, Greene ordered it torn down and rebuilt to his specs. Irksome as it must have been to Coleman and the clients, this fanatical attention to the walls proved right in the long run. A sample of Coleman's unsupervised work can be found in a security wall added years later. It is quite ordinary.

The south-facing bathroom wall is a good example of the attention Greene paid to exterior surfaces. This one-story gable is among the simplest building forms, but Greene turns it into a thing of rugged complexity (photo left). Both corners are slightly battered. They stand out as buttresses from the primary wall plane, and cast narrow vertical shadows that frame a massive arch. The arch (drawing, above left) springs from stone impost blocks that continue in a line roughly in accordance with the top of the window. This line is the only continuous horizontal band in the wall. It

From bedrock to tile roof. **The boulder at the bottom of the photo at right is part of the site's granite. The stonework above it, however, is manmade. Greene chose eroded stones for the first few courses, followed by smaller quarried granite rocks that take over exclusively as the wall becomes well defined. Near the front door (middle right), the intersecting walls blend into one another by way of a small arch. To its left, a limestone scupper is fed by upside-down barrel tiles embedded in the wall to form rain gutters. The roof tiles in the upper right corner have been modified with nippers to soften their edges. Greene had the clay tiles arranged in lines that curve and wiggle as they move up the roof, enhancing the organic colors and lines (top right).**

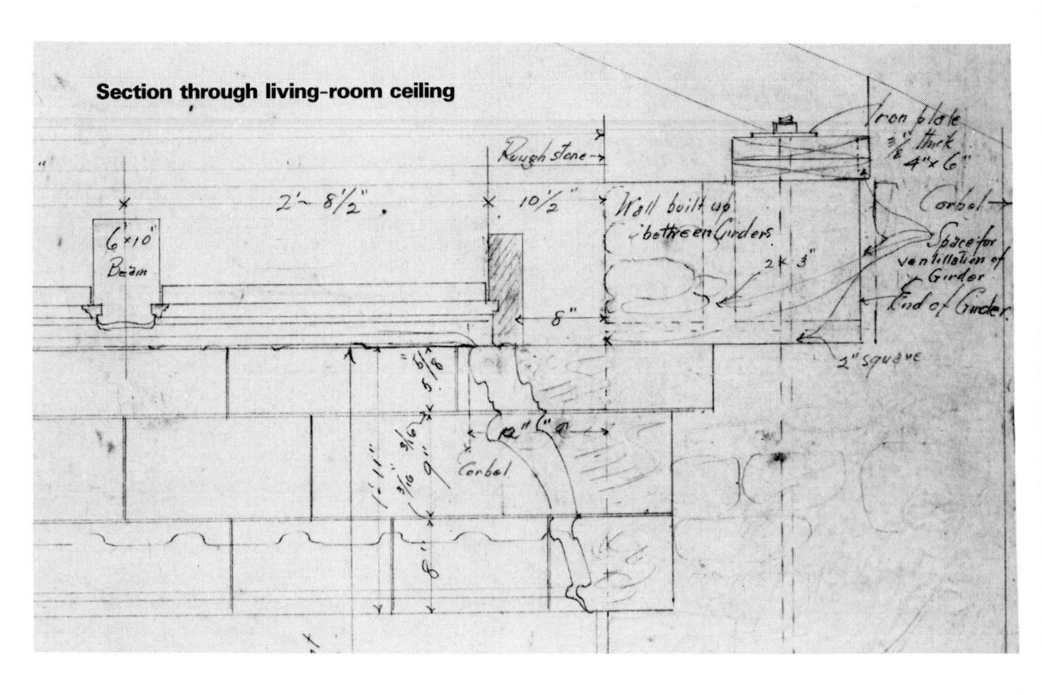

Each corner of the sun room has a recessed bookcase behind steel-banded oak doors, above. The living-room ceiling, top, is a grid of heavy redwood beams, girders and panels that were carved and wire-brushed by Greene, and left unfinished. The section drawing by Greene, top right, shows the construction of the living-room ceiling and the double wall-plate design common throughout the house.

implies the ceiling height, and without it the composition would be uneasy.

To complicate matters even more, the arch tapers—the exterior radius is greater than the interior one, and they originate from different center points on the window's axis. The stonework forms a chamfered reveal 2 ft. across as it moves inward to the window. At the base of the wall is a narrow bench, facing south and sheltered from the prevailing wind.

Greene chose white marble sills and thresholds for the windows and doors. They are bedded on a layer of mortar, and they have weep holes near each end to channel off wind-driven rain. The window frames are teak.

As the stonework surrounding the rough opening for a window or door crept upward, the masons buried 5⁄16-in. anchor bolts in the rock to secure the sash. Greene designed special jambs with a slot for the anchor bolts (drawing, facing page). Once the sash was centered in its opening and resting on its sill, the nuts were tightened and the voids between the sash and the rock filled with mortar. The masons then finished the detail with plaster, which is keyed in place by the back-cut faces on the jambs. Teak plugs hide the anchor bolts. Shutters on every door and window protected the house during the winter, when the James family wasn't in residence.

Greene's daily presence on the job allowed him to make refinements. One of the most intriguing occurs at the intersection of the east wing and the gabled entryway (middle photo, previous page). For most of this juncture, the flat rocks are woven together into a sharp inside corner. But a foot from the eave line the corner suddenly arches over and blends into its neighboring wall. Greene capped this unexpected whimsy with an inspired use of barrel tiles—he had them shingled into the stonework to form rain gutters.

Passion vs. pragmatism—James was a patient man, but not altogether prepared for the experience that unfolded. Greene assured James that it would take two years to build the house. This seemed reasonable, and the men began work on the foundation and retaining walls. On his visit the following year, James was shocked to see that the stone walls were indeed growing, but they were growing down the cliff, as Greene anchored the house, visually as well as structurally, to the site.

By 1919, the major foundation work was done. The frequent correspondence that passed between Greene and James reveals that the architect's interest in the project had grown into a passion. A drawing he sent to James in April 1919 shows the project to date. Its simplicity is unusual for Greene, and it shows that the floor plan was unresolved. In the accompanying letter, Greene refers to retaining walls to the north, fill for the courtyard and hand-carved marble mantelpieces. Two years into it, and the foundation was nearing completion about the same time the marble details were being carved.

Sensing the scope of his masterpiece, Greene produced reams of drawings. No decorative detail seems to have escaped his attention, yet the archive (Documents Collection, College of Environmental Design, University of California at Berkeley) doesn't contain what might be called working drawings. Structural solutions must have been worked out on the site. Without his brother's practical hand to guide him, Charles Greene did what he knew best—envisioned perfection.

By 1920, James had realized that Greene didn't want the project to end. The architect was finicky beyond belief, and his obsession was going to drive James to financial ruin. The absentee client worried about idle crews waiting for Greene to make up his mind, and James finally told the architect to wind things up. Ironically, James' decision included the veto of his upstairs study.

Drawings of the house confirm James' suspicions. When Greene drew a bookcase, he sometimes drew the books too, with details of their spine decorations. When he made elevation drawings of the sun-room bookcase doors, he did separate full-scale drawings of

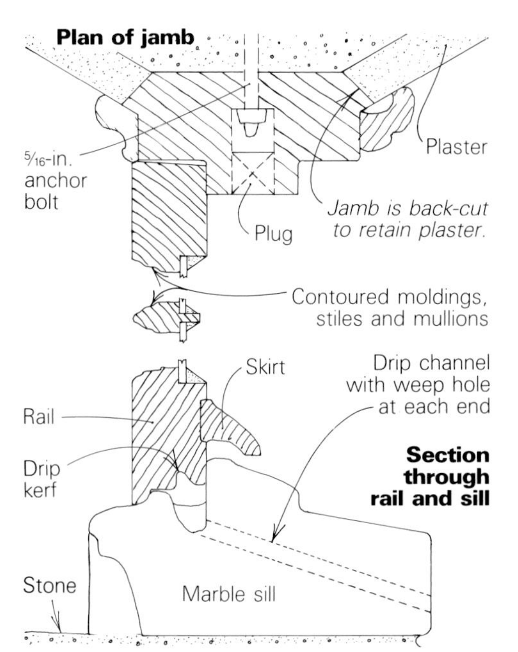

the hardware, showing the placement of the door pulls, complete with hammer marks on the rivets. No wonder James began to doubt the house would ever be completed.

A tile top—Greene chose Spanish barrel tiles for the roof. Their permanence, texture and color are perfectly in keeping with the rest of the house, and Greene used them to form tiny gables over the chimneys and buttresses, creating a village of cascading roofs. The tiles are from Gladding McBean (P.O. Box 97, Lincoln, Calif. 95648), a company that still produces terra-cotta products. Most of the tiles are shades of rusty red, but every now and then the architect slipped in a green, salt-glazed tile for variety. As he did during the placement of the stones in the walls, Greene stayed on Coleman's case throughout the roof work. Greene was after a wavy, random pattern in the tiles, so he directed that the tiles be positioned in curving rows (top photo, p. 55). Along some rake walls, the factory edges of the tiles have been nibbled with tile-biters to texture their edges. At the ridge, the cap tiles rise and fall on a bed of mortar and shards. This saurian ridge cap wanders along the horizon as though out for a swim in a calm pond.

By the time the roof went on in 1920, the house was a popular attraction, and the locals would stop by to check on its progress. They would frequently alert Coleman to the unkempt rows of tile, thinking they'd discovered a mistake. Coleman had several replies, including, "This is the way the architect wants it. . . He's a crazy person." Or he would jest, "The clients are crooked, and so is the roof."

A crawl through the attic reveals the roof's underpinnings—full 2x8 rafters at a 5-in-12 slope, 12 in. on center, with diagonal braces under every other rafter pair. This arrangement has carried the barrel tiles, which weigh

Seen from the sun-room library, the entry door is in the center of the photo. Like the rest of the rooms, the sun-room walls are finished with a layer of unpainted, sand-finished plaster.

Charles Sumner Greene

Charles Greene was born in October of 1868, near Cincinnati, Ohio. His brother Henry was born 15 months later. Their father was a doctor specializing in the treatment of catarrh, and the boys enjoyed a comfortable middle-class upbringing. They attended a high school with a manual-arts program, where they were talented and enthusiastic tool users.

After graduation from high school, the brothers studied architecture at MIT. They each completed a two-year program and went to work in separate Boston offices. Henry found the work to his liking, while Charles found it tedious. Charles was more interested in history, painting and philosophy, and he contemplated quitting architecture.

In 1893, the brothers visited their parents, who had moved to Pasadena, Calif. Charles and Henry were taken by the mild climate, lifestyle and the informal architecture, and they decided to settle there and open an office.

The Greenes designed houses for ten years in a variety of styles, including Colonial, Mission, and the Shingle Style popularized by H. H. Richardson. Gradually, they discarded the applied ornament of accepted fashions in favor of articulated structural elements—steel-banded scarf joints, wedged tenons and doors with shapely rails and stiles. Their houses embraced the landscape with wide eaves, supported by rounded rafters and brackets. As their style emerged, no detail was too small for inclusion in the overall scheme. This dedication culminated in their "ultimate bungalows"—vast landbound luxury liners that rise from the center of their stadium-sized lawns. They are mostly wood—the finest mahogany, oak, redwood and teak—and the Greenes filled them with furniture, light fixtures, leaded glass and rugs of their own design.

The Greenes also worked in more modest veins. Their Bandini house in 1903 transformed the traditional Spanish courtyard home, always made of adobe, into a board-and-batten forerunner of the California ranch house.

During their partnership years, the brothers had well-defined roles. Henry's even temperament, engineering talent and organizational skills held the firm together, while Charles' artistic vision and insistent nature shaped the firm's designs. Charles, a diminutive man with a page-boy haircut and a voice that somehow managed to be both shrill and barely audible, pushed their contractors to the limits of their talents.

Their glory years in Pasadena came to an end as World War I began. Pasadena had grown from a vacation spot to a self-conscious little city, and revival styles from Europe became the fashion. In 1916, Charles moved to Carmel for a change of scene, and built a small studio where he sporadically pursued architecture, writing and Eastern mysticism. One two-page spread in his James house construction notebook reveals the breadth of Charles' interests. It contains the James house roof plan, formulas for removing paint from clothing and calculating material for an oak floor, an abrasive schedule for marble polishing, a design for a clothesline prop, and a recipe for Roquefort dressing.

Charles Greene died in 1957. Twenty years later, the influence of the Greene brothers exerted itself once again when Randell Makinson's study of their work, *Greene and Greene, Architecture as Fine Art* (Peregrine Smith, Inc., Box 667, 1877 Gentile St., Layton, Utah 84041; $24.95), hit the bookstores. Makinson's book came at a time when a lot of architects and builders were ready for artistic inspiration and moral support, and once again the Greenes' achievements and high standards found an appreciative audience. —*C. M.*

1,300 lb. per square, with no sign of deflection. The sheathing is 1x fir, covered with a layer of 30-lb. asphalt-impregnated felt.

Incredibly, the roof was done twice. The first time it was laid with mortar pads under the tiles along the eaves, followed by tiles that were wired, but not mortared in place. Unfortunately, maintenance trips proved unavoidable, and walking on the roof would break the tiles. So 12 years later, the tiles were pulled up and individually mortared in place.

A soft interior—The inside of the James house presents a powerful contrast to the angular surfaces of the exterior. The transition begins at the windows and doors, where the plaster walls squeeze past them to envelope the arched openings. This smooth surface signals indoors, and softens the house to the eye and to the inevitable brush with a hand.

The entry is on the courtyard side of the house, and its generous circular hallway allows a first glimpse of the ocean through a series of arches. In the original plan, the hall was to have been continued to the sea side of the house, but James decided to use this space as a library—probably after it became apparent that realizing the original plan would take too long and cost too much. Called the sun room in Greene's plans, this space (photo previous page) is typical of the formality James desired. The wall between the library and the entry is a trio of arches, symmetrically flanked by oak doors that open to reveal bookcases. The bookcase doors (bottom photo, p. 56) are pure Greene and Greene, with hand-wrought banding and pulls, and a light coat of green stain to soften the red oak.

A limestone coving takes the walls into the plastered ceiling. Greene directed the stonecutters to shape the crown pieces with studied irregularities, and the coving appears to have the texture of adze marks. This bumpy surface is quietly apparent under even light, but a fire in the fireplace quickly changes the effect. Then the hollows create shadows that rise and fall with a flickering pulse.

In the living-room ceiling, Greene made a grid of 14x14 redwood girders that are intersected by 6x10 beams. The girders extend far beyond the limestone corbels that appear to support them (drawing, p. 56). Both girders and beams have grooves cut into their sides to carry ledgers for the redwood infill panels. And what extraordinary panels they are—the largest are single boards, 1½ in. thick, 6 ft. long and 46 in. wide (top photo, p. 56). Greene textured the panels with quick, hatch-like gouge strokes a few degrees off the grain direction, and relief-carved the bottoms of the girders and beams with simple, wavy lines.

Plans that were never used—Greene's drawings for the James house show many furnishings and and decorations that were never built. Among them are elaborate tables, chairs and a desk. All were to be white oak, with ebony accents. The furniture is in the vaguely oriental style that the brothers used in earlier homes, with the addition of relief-carved birds and cloud forms. In a playful moment, Greene sketched a bespectacled owl watching over the books in the library.

The Jameses decided against the furniture, saying they thought it lacked the visual weight demanded by the house. But chances are, after five years of construction, the clients and the architect needed to get away from each other, and that meant leaving some things unfinished. James, the patient, organized businessman, had coaxed the fading genius through a masterwork. It was time for him to enjoy his house, and that he did. His son Daniel recalls James taking special delight in eating his breakfast on the terrace, and waving to the tourists who ogled him through telescopes from the next point down the coast.

Library addition—In the late 1930s, James decided to add a proper study to the house. He asked Greene for ideas, and they mulled over various schemes, including one that would have transformed the circular bathroom into a library. Eventually they followed a suggestion by Daniel to rework a space under the maid's room. James approved Greene's plans, which included a wall of teak bookcases, a marble-trimmed fireplace, and a small writing alcove surrounded by columns topped with bush-hammered limestone capitals.

Work began in 1940, and you can trace its fitful progression in the correspondence between Greene, James, the woodworker and the stonecarver. Greene was in his 70s by this time, and his letters reveal his growing frailty, and the pressure he felt from his client. His drawings show his preoccupation with detail to be undiminished, but they became smaller, layered with too many lines for their scale and hard to understand.

Meanwhile, wartime shortages eliminated some materials, such as teak, from the project, and James dismissed the stonecarver, writing to Greene that "life is too short to bother with a man as completely inefficient, slow-thinking and unbusinesslike. . . ." Prophetic words they turned out to be—James died in 1944, with the study just short of completion. All work was halted until 12 years later, when Elizabeth Gordon, editor of *House Beautiful,* persuaded Mrs. James to finish the room. A part of the cost was assumed by the magazine's advertisers, who used the room to showcase their wares.

Although he never used his new library, James wrote a great deal during his years in the house. Play after play came from his pen, but all fell just short of professional production. They were infused with an Edwardian romanticism that was no longer fashionable on Broadway. So James' legacy was not literary, but architectural; the house he commissioned to inspire his writing turned out to be his most enduring monument. □

Original drawings p. 52 left and p. 56 courtesy U.C. Documents Collection; site plan p. 52 adapted from Greene and Greene, Architecture as Fine Art; *1923 photo p. 52 courtesy Slevin Collection, Bancroft Library, U.C. Berkeley.*

A View of the Redwoods

Owner-builders strike a balance between fastidious detail and raising kids

by Charles Miller

Northern California's Mendocino coastline is a craggy strip of headlands, beaches and tidal pools that draws visitors from all over the world. Quaint inns and rustic farmhouses that have been converted into bed-and-breakfast retreats are politely scattered along the Coast Highway. Sitting in their dining rooms you can munch on a bran muffin and sip coffee while you watch the salmon boats and the grey whales go by.

But if you go outside on the Mendocino bluffs, be sure to take your mittens and wool cap. A cold wind blows in off the ocean, so most seaside residents spend a lot of their time indoors. The picture windows that were carefully oriented to frame the best views turn cloudy from salt encrustations. An offshore fogbank hangs over the coastline, plunging the cliffs into a grey murk. Meanwhile, a mile inland the sun shines warmly on the redwood forests.

Terry and Cecily Klingman had been around Mendocino long enough to be familiar with its climate. While some coastal locals welcome the wind and the fog, the Klingmans wanted to build their house where they could spend a lot of time outdoors with their two young children, Nina and David. They concentrated their search for land a few miles inland, and settled on a 10-acre parcel just 10 minutes from town.

The Klingmans both grew up in the East, in homes they characterize as inward and formal. When they moved to the forests of Mendocino, they wanted to leave architectural formality behind and build a house that opened onto the woods with big windows, balconies, decks and plenty of skylights. Since they were living in a redwood forest they wanted to use redwood for siding, beams and interior trim.

Terry had had some building experience in his native Michigan, so he planned to be the contractor for the house. He had a lot of talented builder friends from his home state who were eager to move west to help build the house. Since Mendocino is a haven for accomplished woodworkers, finding qualified people to work on the house wouldn't be a problem.

An octagonal plan—While the house would be an owner-builder project, the Klingmans needed a designer to give form to their ideas. They chose Michael Leventhal, a Mendocino architect known primarily as a modernist, and asked him to design a house that would take in as many of the surrounding views as possible. Initial thoughts of a round house were shelved—too complicated to build. Instead, Leventhal proposed an octagonal plan (drawing, p. 61), with windows that look out in all directions.

The house has two levels. On the lower level, the plan is essentially cruciform, with what amounts to five large rooms. The living room, dining room and kitchen are aligned on the east-west axis. An office occupies the north corner, and the southern corner is devoted to an airy sunroom with a translucent fiberglass roof (photo below). A circular stair leads to the second floor, where the plan deviates from the octagon. Here three bedrooms are massed on the north and east sides of the house.

The Klingman house is on the edge of a meadow that backs up to a redwood forest. In the this view from the south, the sunroom is the predominant space on the lower level. Upstairs, a bay window extends from one of the bedrooms. On the right, solar hot-water panels mounted on the garage roof peek over a tree.

The kitchen, dining room and living room are uninterrupted by interior walls. They form a continuous space a little over 50 ft. long. The living room (photo next page) is three steps down from the rest of the first floor, and its ceiling is higher than that of the dining room. This room has a lofty, symmetrical feel to it. Along its ridge is a narrow skylight that is on the same centerline as the monumental cast-concrete fireplace. Windows flank the fireplace, framing it with a dense thicket of green treetops. The room's strict geometry befits its role as the only formal space for grown-ups. The rest of the house was designed with children in mind.

Making space for children—Like everybody else, kids want to hang out in the kitchen. But until they learn how to dodge distracted cooks, they can get into trouble underfoot. Knowing this, the Klingmans requested a designated children's space adjacent to the kitchen, where they could play without endangering themselves.

Leventhal's solution is a triangular alcove on the kitchen's southern edge (top right photo, p. 61). It has a low counter where several youngsters can draw at one time. A big window opens onto the sunroom to let in heat and light, and a nearby closet and bookshelf provide plenty of storage. Although the alcove is finished with

On the right, a circular stair carried by a polished concrete column leads to the upstairs bedrooms. To the left, a cast-concrete fireplace is the visual anchor in the living room. Redwood beams and rafters carry the weight of the house, and the living-room ceiling is made of split redwood stakes normally used for fences.

An alcove for the children (above) is located on the southern edge of the kitchen, where it gets heat and light from the adjacent sunroom. Windows between the counter and the overhead cabinets illuminate the kitchen's work surface (left). The island cabinet in the foreground contains the cooktop, and space for firewood for the nearby woodstove. The doors on the left corner of the island conceal a pair of bins on extension hardware. They hold kindling and newspapers.

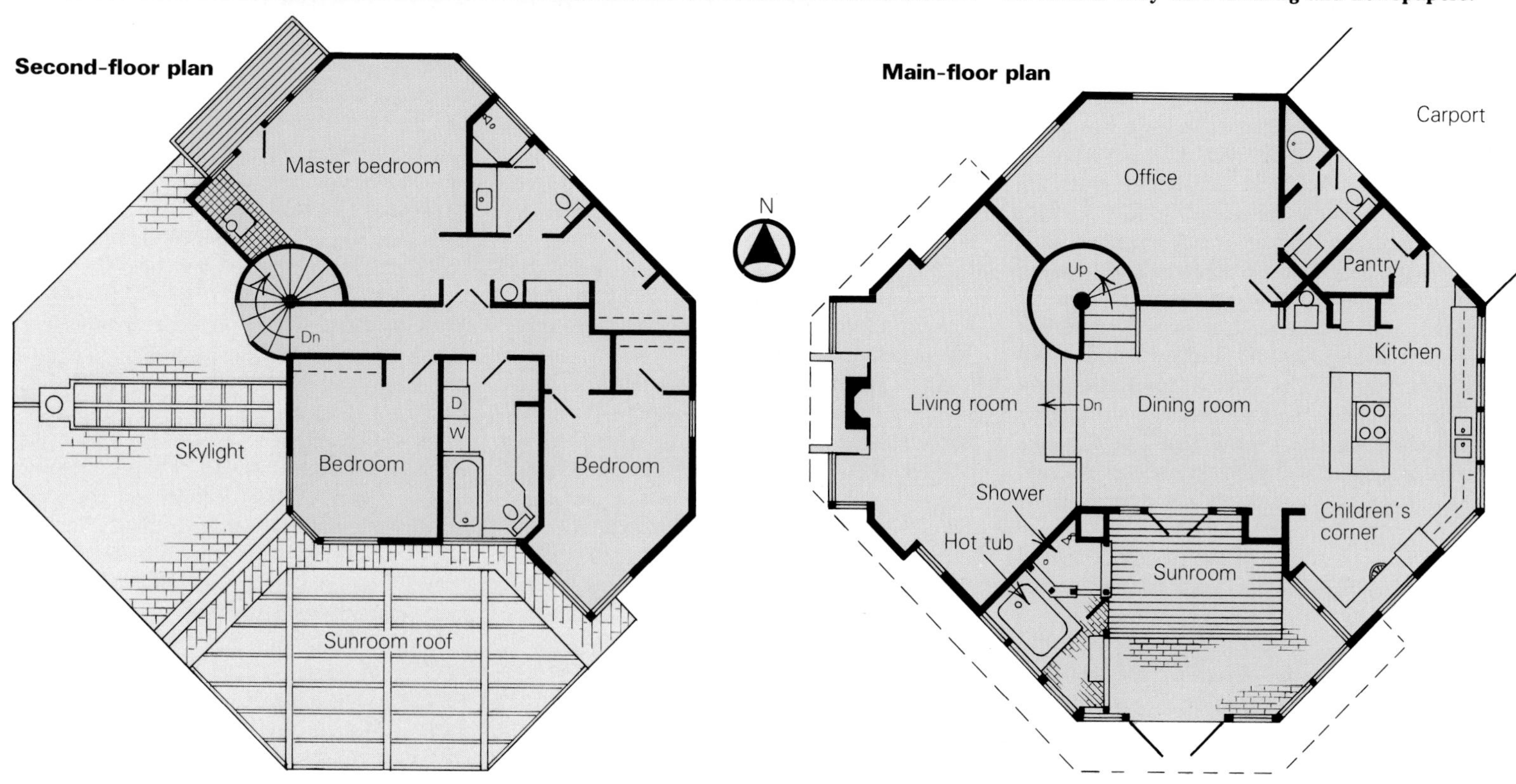

A 4-ft. square skylight illuminates the children's bathroom on the second floor (left). When the sun goes down, the Craftsman-style vanity light by Mendocino woodworker John Birchard goes on. Retractable steps in front of the sink boost short people to the right height. A wide shelf around the tub holds tub toys (above), and the dark tile frames the bottom of a mural by William Stoneham.

hardwood flooring, the Klingmans had a carpet installed for David's toddler years. A low fence along its angled border with the kitchen serves as a baby gate. When the children get older, the space will become a breakfast nook.

Although the kitchen wasn't intended as a command post, its line of low windows between the counter and the upper cabinents (top left photo, previous page) has proved useful for keeping track of the children when they are playing outside. The low windows also serve their intended purpose. Leventhal knew that the Klingmans wanted to use cobalt blue tiles as a color accent above the stainless-steel counter, but he was concerned that the deep blue tile would make the counters too dark to be comfortable work surfaces. The windows lighten things up, and make the upper cabinets appear to hover above the stainless steel.

A children's bath—Anybody with an infant needs a good place to change diapers, and the children's bathroom in the Klingman house has one. A long, 36-in. high counter stretches along one wall of the bathroom (photo above left). The sink is at one end, and it is large enough to bathe a child who is too young for the tub. The sink is equipped with the kind of hand-held sprayer normally found on kitchen sinks. Here it is used for washing hair. The counter space to the left of the sink is the diaper-changing station. A sturdy railing affixed to the cabinet face frame extends above the counter, preventing an exuberant baby from rolling off.

At 36 in., the counter is tall enough to be a comfortable work height for adults—but what about a 4-ft. person trying to use the sink? The solution here is a retractable two-tread step, complete with handrails. The treads are mounted on a plywood carriage that rides on heavy-duty drawer hardware. The carriage just kisses the floor when the step is pulled out, so the weight of someone standing on the step bears directly on the tile floor.

The cabinets look like standard louvered doors, but they are actually drawer faces for laundry bins—one for clean diapers and related baby gear, the other for dirty clothes. Within arm's reach is the washing machine and dryer, and the linen closet is in the corner.

A tiled shelf to the right of the sink continues around the room, eventually bordering the bathtub. The shelf holds the kids' tub toys, and the dark blue tile is a visual anchor for the mural above the tub (photo above right).

Wall murals—The Klingmans commissioned Mendocino artist William Stoneham to paint several walls in the house. In the kids' bath, the theme is water, and Stoneham used his polymer acrylics to conjure up a seascape that reveals what goes on below as well as above the waterline. A gallery of bass, starfish and crabs peer down at the bathers. A porpoise with Nina as a passenger heads west, while on the opposite wall, Christopher Columbus' boat of the same name approaches the mirror.

In David's room, the theme is medieval fantasy. In one corner, Stoneham painted a castle (photo facing page, left) rising from a narrow loch, complete with distant sea serpent. A friendly giant sits on the path to the castle, and

Stoneham's mural in David Klingman's bedroom centers on the castle scene at left. Above David's bed, a bear and his elf go for a balloon ride.

an elf drifts by on a bubble blown by a dragon across the room. When the bubble pops, the elf takes to riding a bear along the top of the window casings. Eventually the elf coaxes the bear into taking the plunge off the window trim, and holding a balloon, they float down toward David's bed (photo above right).

Woodwork—The trim detail abandoned by the bear is typical of the trim treatment found throughout the house. A 1x4 vertical casing with an ogee profile on the window side abuts a horizontal 1x2 at the header. The 1x2 extends about 1 in. beyond the line of the vertical casing. A 1x6 header casing runs along the top of the 1x2. Its square-cut end continues the line of the vertical casing. The overall effect is one of rich color coupled with some judicious lines of highlight and shadow. Since there are no miters, the detail was pretty straightforward to install.

Because this detail runs throughout the house (along with lots of 1x wainscoting), Klingman needed a lot of dry, clear redwood. The going rate for dry vertical-grain stock at local retail outlets at this writing was about $1.50 per bd. ft. But Klingman found that local yards often had mill runs of clear redwood that were slightly under standard dimensions. These oddball lots, along with some that included sapwood streaks, could be had for $.60 to $.70 a bd. ft. Klingman bought piles of the stuff and sorted the pieces by color, grain and dimension. Then the crew ripped the stock to the appropriate width and used routers to mill decorative edges.

Before the trim pieces left the workshop, Klingman sprayed them with a couple of coats of lacquer. This finish kept the trim from falling prey to dirty fingerprints and spilled coffee during installation. But lacquer has a drawback as a trim finish. When dented, lacquer will lift away from wood, leaving a milky white halo around the ding. With kids running about the house, a lot of creases and dents started showing up in the trim, especially the door casings.

The solution to the lifted lacquer turned out to be Flecto Varathane Plastic Oil. Rubbed into a blemish with a cloth, the oil penetrates under the lacquer to rebond it with the wood, and also adds a luster to the woodwork as it accumulates the scars of use.

In the kitchen, the cabinets are made of cherry. The counters, floors and cabinet pulls are ifilele (rhymes with ukelele), a dense, reddish wood from the western Pacific that resembles teak in appearance and utility. Cabinetmaker Gary Church built the island counter in the kitchen out of cherry, and used ifilele for the pulls.

The evolving house—A two-story garage was the first structure to go up when the project began. The downstairs was the woodshop, and the upstairs served as the finishing room for lacquer work. Now the upstairs is a guest house, and a roof connects the garage with the main house to form a large carport.

The sunroom was initially intended as the formal entry to the house, but everybody comes in through the door to the carport. So the Klingmans added a hot tub and a shower to the corner of the sunroom. The tub is made of ferrocement, with a deep end for adults and a shallow end for Nina and David. ☐

Facade Roof

Structure is ornament in this solar-heated, steep-site design

by M. Dean Jones

From the street, your first view of this house is of the roof (photo facing page). Its shingled surface is symmetrically interrupted by skylights, and a small, four-gabled observation tower rises at the center, its roof planes hovering over those of the main roof. The rafter tails and open soffits that extend beyond the eaves are the first suggestion of the exposed structural system that is integral to the overall design.

After descending about 20 ft., you approach the main level of the residence by crossing a small bridge (drawing, below). The end of this walkway widens to form an entry porch beneath a skylit roof. Here the detailing of the roof system comes into full view. This interplay of wood structural members crossing and joining each other continues inside the house.

As an architect, the decision to celebrate the carpenter's art in Mill Valley, Calif., was not a difficult one. Craftsman-style architecture had strong beginnings in the San Francisco Bay Area around the turn of the century. But in addition to creating a modern expression of Craftsman-style architecture, I also wanted to design a house that would meet the modern-day needs of my clients, Art and Marianne Quasha. What we started with was not an empty lot, but a difficult house.

Problem house, beautiful view—When Art and Marianne Quasha bought their Mill Valley property in 1978, it was the site and not the house that attracted them. The lot slopes down steeply from the street, facing east with a dramatic 200° panorama (photo right). Included in this view are the San Francisco skyline and Mount Tamalpais, a 2,586-ft. landmark that dominates the topography of southern Marin County. If properly designed, a house on this site could take advantage of solar-heating potential and at the same time capture this tremendous vista.

The house that came with this dramatic site had definite problems. A one-story, wood-frame structure built some 50 years ago, it was an artist's delight with lots of north-facing, single-glazed windows. But it was a conservationist's nightmare. There was neither south-facing glass nor insulation. The flat-roofed structure also had dry rot and mildew problems, as well as substantial air infiltration through hundreds of cracks. Moisture in the wood, combined with slab-on-grade construction, had made the house an ideal habitat for termites. The final disadvantage was one of inconvenience: the building pad was located 84 steps below street level.

A fresh start—The Quashas lived with the cold, creaky old house for four years, carrying groceries down many steps and coming back up again day after day. Having to carry their infant son didn't make things any easier. They plotted ways to change the old place, talking with friends, reading magazines and consulting with architects and builders about the transformation they desired. The first thing they did was to have a two-car garage built at street level with an attached guest house that extended down below the street, following the slope of the lot.

The Quashas contacted me because of the experience I've had in designing energy-efficient houses on hillside sites in many parts of the country. We talked at length about the house they wanted, and studied solar patterns at the site. Then we made our major decision—to wreck the existing house, break up the slab and start with a new structure.

After the demolition took place, all that remained were some old beams and studs that were sawn into firewood for the three wood-burning stoves that would be installed in the new home. Working together, we developed a design for the house that reflects the region's architectural heritage and also develops the potential of their steep site. By designing the main living level at an elevation halfway between the original house elevation and street level, the distance from garage to living space was cut in half. We also designed a new entry path (still to be built) that conforms to the natural slope of the land. This reduces the perceived vertical distance between house and street.

Even with a new entry and higher house elevation, it was clear that the roof of the new house would be its most prominent feature from the street. Unlike most roofs, this one would have to express strongly the style that would be part of the interior space. From our design studies we developed a roof that would answer our many requirements. It serves the functional need of protection but also brings in light and solar heat. The observation tower and its sun deck offer wonderful views of the Bay Area. When open, the triple-glazed skylight at the tower's peak encourages convective cooling throughout the house. And a glass floor in the tower brings light down to the main level of the living space (photo next page, left).

To achieve the level of detail that my design called for, I had to depend on a skillful, cooperative contractor. Charles Melin, who has built at least 15 of my designs, again came through with an excellent job on this house.

Four levels—The house has a total of 3,648 sq. ft. of living space, not including its observation tower and numerous decks (see the floor plans and section on the following pages). Just inside the entry porch, a small foyer eases the interior transition and also acts as an airlock. From here, you step into a hallway we call the gallery. On the left, there's a study with a half-bath. On the right, a wall of bookshelves leads you into the living room and dining room, which span the full 32-ft. width of the house (photo next page, right). The gallery is also the location

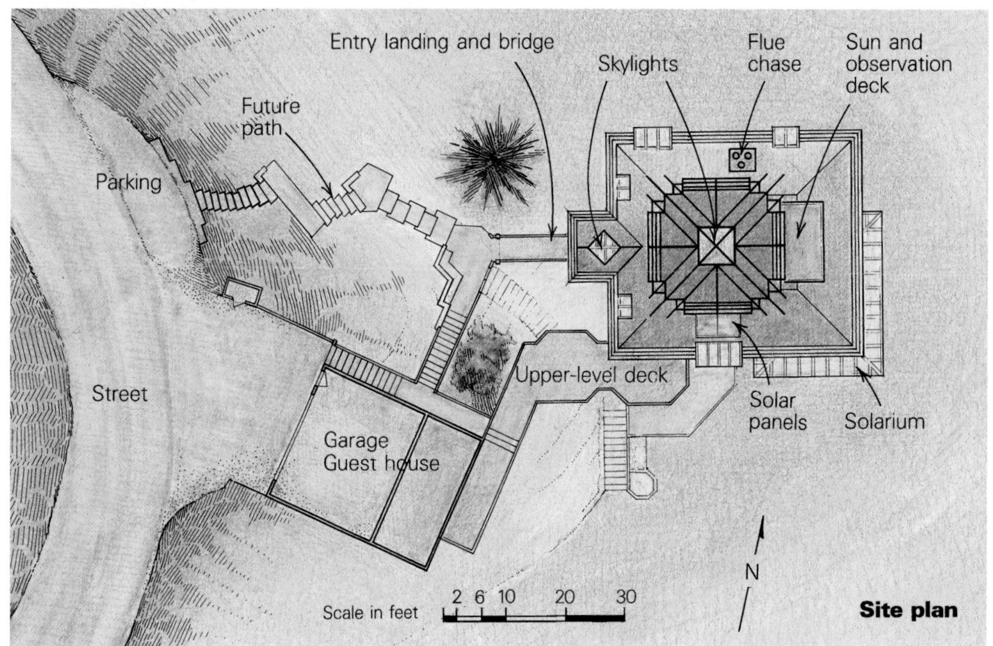

Facing page: Because of the view from the street, the roof became the dominant design element in this house, which occupies a steep site in Mill Valley, Calif. The intricate eave detailing suggests the display of structural woodwork that continues inside the house. An observation tower topped with an operable skylight penetrates the main roof. Above right, lush natural surroundings and a panoramic view are the site's chief attractions.

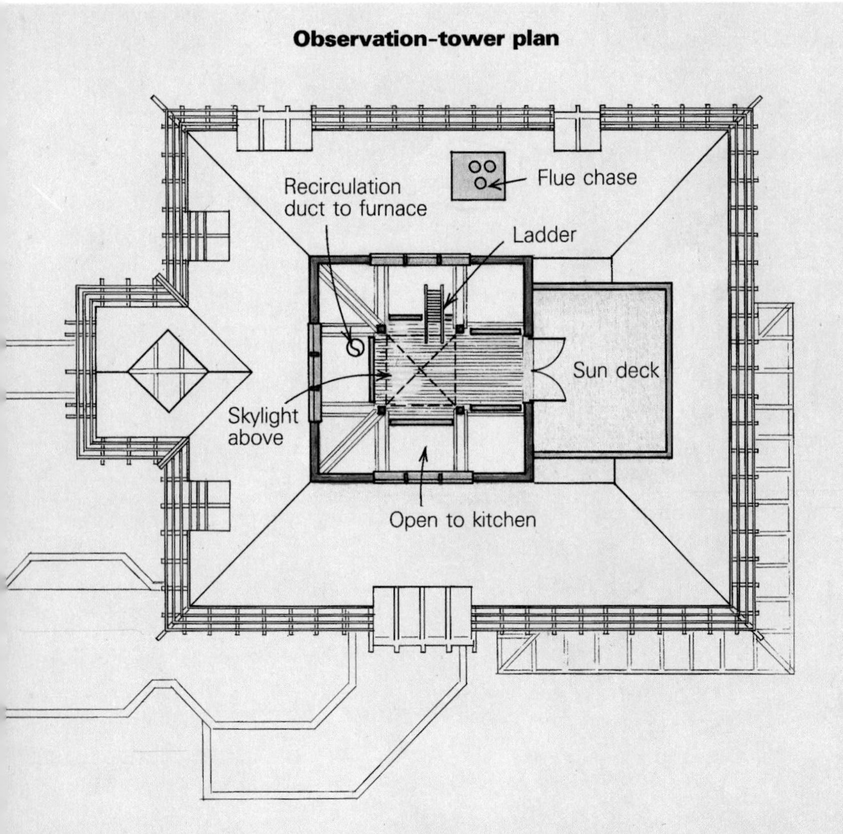
Observation-tower plan
Flue chase
Recirculation duct to furnace
Ladder
Sun deck
Skylight above
Open to kitchen

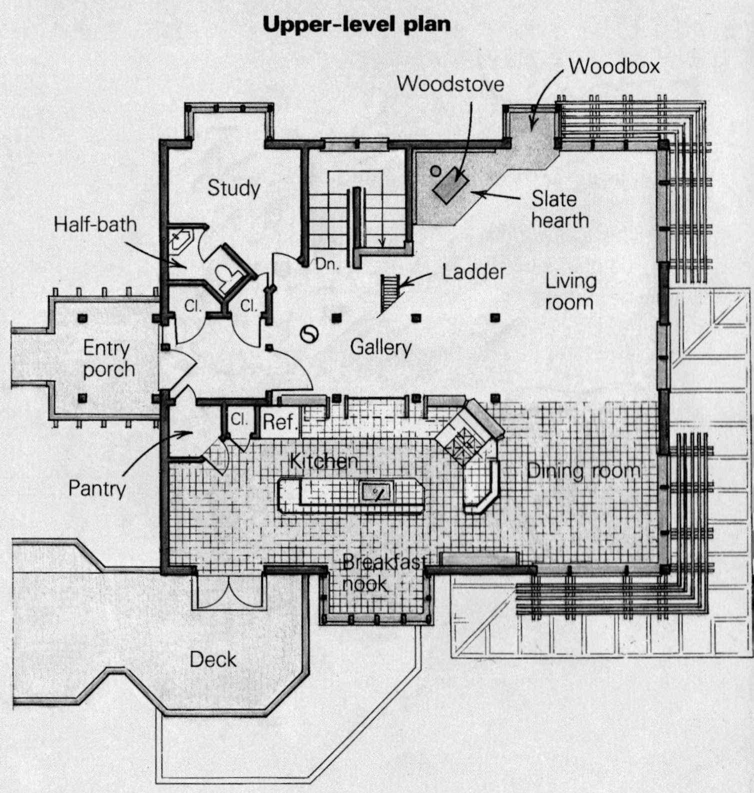
Upper-level plan
Woodbox
Woodstove
Study
Slate hearth
Half-bath
Dn.
Ladder
Cl.
Cl.
Living room
Entry porch
Gallery
Cl.
Ref.
Kitchen
Pantry
Dining room
Breakfast nook
Deck

The carpenter's art. **The crossing and joining of structural members is here celebrated rather than hidden. The glass floor of the observation tower (facing page, left) brings natural light into the depths of the house. Access is by the ladder on the right. An overlook on one side of the tower offers a view of the kitchen, above. Light-colored tile countertops contrast with the dark tile floor and redwood shelves. Beyond the structural geometry of posts and beams (facing page, right), a bookcase wall leads back toward the entry foyer.**

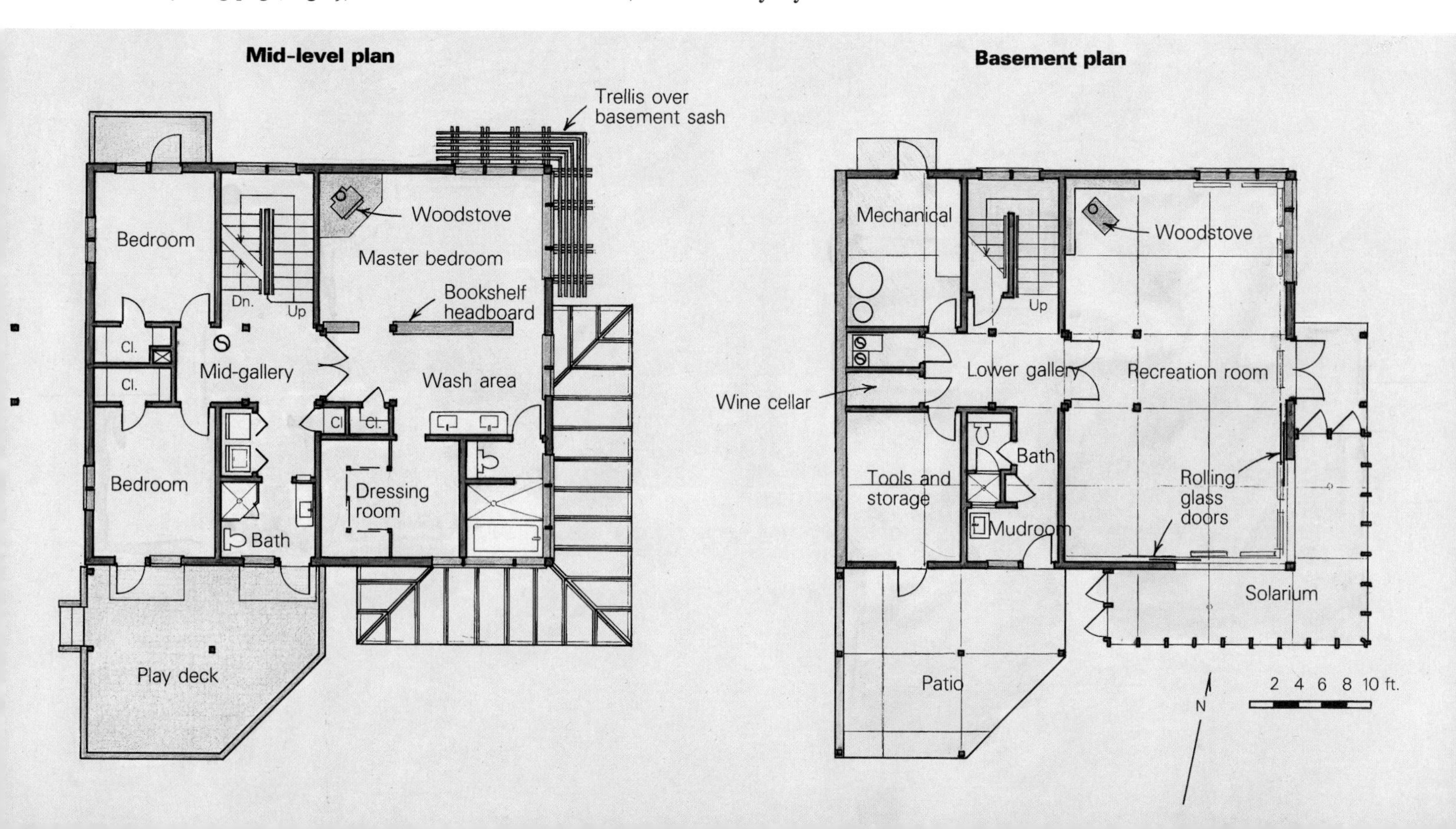

An outdoor focus. **The upper levels of the house have decks, and the ground floor has a solarium that wraps around the southeast corner.**

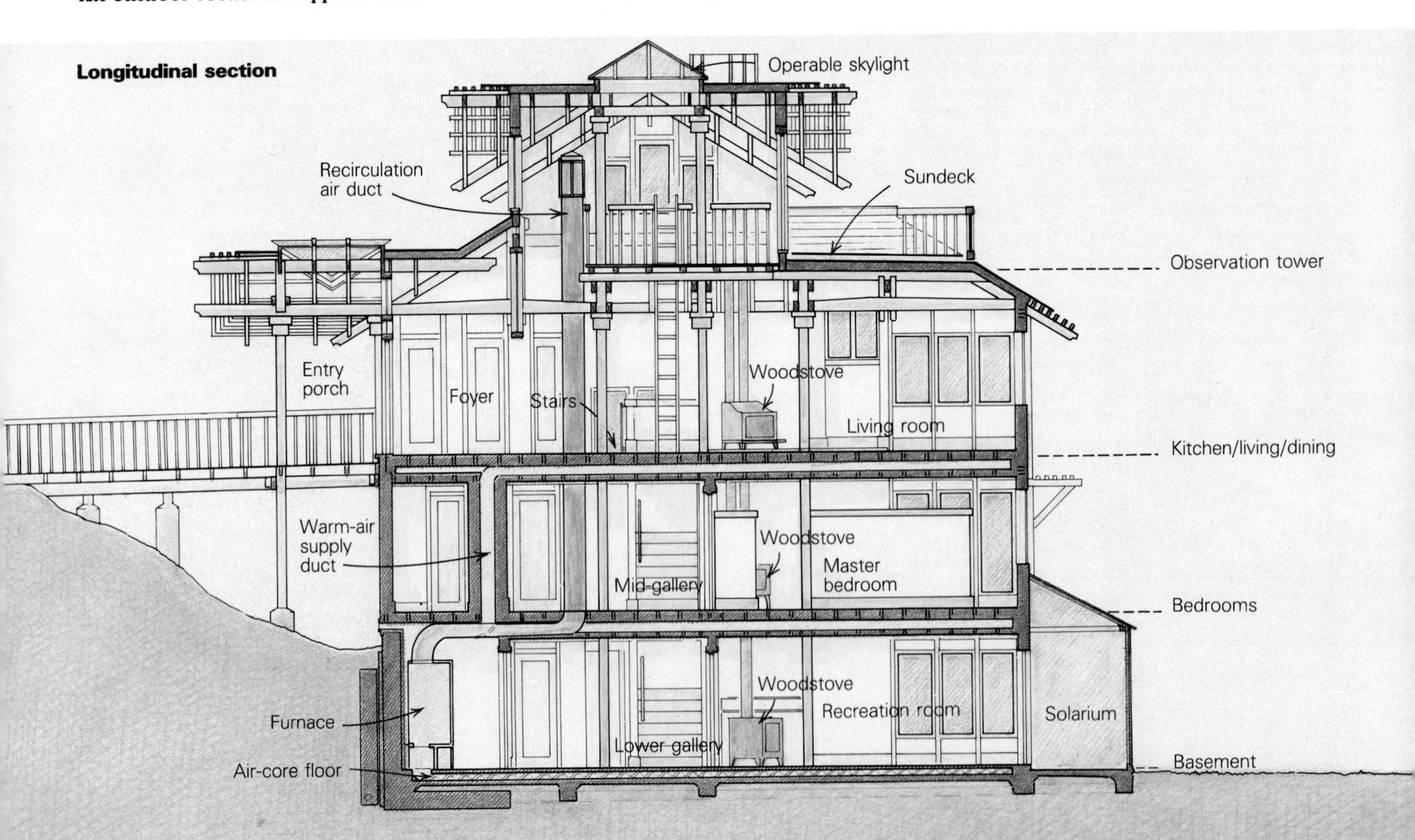

of the "captain's ladder" that takes you up to the observation tower. Here you can gaze outside to distant views, go out to the sun deck, or look down into the kitchen.

I designed the kitchen to be not only an efficient work space but also a social space so the chef would be able to talk with family and guests. A small commercial gas stove and griddle, with Marianne Quasha at the controls, is the center of many dinnertime discussions. There's a broom closet and a walk-in pantry. A long island forms one side of a galley-type layout. The cabinets in the kitchen and in the rest of the house were made from red cedar boards by Kenny Goettsch.

The south side of this level has a thin-mass floor of ceramic tile on a mortar bed. This floor area is heated by direct gain on sunny winter days and radiates this heat throughout the night. Two destratification fans ensure quiet, constant air circulation during heating or cooling periods.

There's additional thermal mass on this level in the form of mortared slate surrounding the woodstove. The slate also lines a projecting bay that serves as a wood-storage box adjacent to the stove. Firewood can be raised from the yard to this level by a pulley system that runs outside the house.

Downstairs, a large landing provides access to the bedrooms. Two children's bedrooms share a bathroom, and each room has its own deck. The master bedroom is actually a suite divided into four main areas. There's a vestibule/vanity as you enter the suite. The bed has a bookshelf/divider headboard and is situated to take in distant views to the north and east. A small woodstove occupies one corner of the room, and in the southeast corner, there's a bathroom that includes an open, two-person shower and a Jacuzzi tub. This area also has a thin-mass floor and walls of stoneware ceramic set in mortar.

The lowest level of the house brings you down to the yard, which can also be reached by stairways from several upper-level decks. This level contains a mechanical room and shop as well as a large storage room, a bathroom, a mudroom with a large sink for washing garden vegetables, a wine cellar and a recreation room. The most dramatic feature, however, is the solarium that fills the southeast corner (photo, facing page). In the summer, when solar heat isn't needed, the sunspace can be isolated from the rest of the house by sliding a pair of large glass doors across the recreation-room opening. These doors were custom made and move on standard barn-door hardware. With the solarium closed off, excess heat is vented naturally outside through doors in both solarium end walls.

Attention to detail. **Custom-made lighting fixtures like the ones in the photos above incorporate leaded and stained glass and continue the Craftsman motif. In the solarium, top, large glass doors roll open and shut on barn-door hardware. The solarium can thus be isolated from the rest of the house, allowing excess heat to escape out side doors.**

Easy to heat—The solarium (photo top right) is a key part of an integrated whole-house heating system. The floor slab for the adjacent recreation room is an air-core floor of concrete that was poured around interconnected stamped-steel channels. The air-core slab (468 sq. ft.) stores direct gain from the sun but can also be "charged" with heat that rises naturally to the top of the house. This is done by using the fan on the forced-air furnace, which pulls warm air down through a recirculation duct to the furnace plenum. When the warm air reaches the plenum, thermostats in different parts of the house tell the warm air where to go (if it's not warm enough, it can first be heated by the gas furnace). If there is a demand in some area of the home, the warm air is sent there. If there is no demand, a motorized damper closes off the access to the duct system and the fan sends the warm air into the air-core floor. Perimeter grilles around the recreation room return the air to the sunspace, where its convective loop through the house can begin again (drawing, facing page).

Once the available heat from the air has been given up to the floor slab, this thermal mass becomes a constant source of radiant heat below the main living spaces of the home. When it's been charged with heat, the air-core floor has a lag time of two to three days. This flywheel effect applies whether the mass is heated by direct gain from the sun or by warm air from the rest of the house.

The air-core slab has R-13 rigid insulation beneath it, while the walls and roof areas are insulated with R-19 and R-30 fiberglass, respectively. Only during long periods of cold, cloudy weather is there demand for additional heat. Alternatively, the Quashas can fire up a large Cawley-Lemay woodstove located in the northwest corner of the recreation room. This stove can heat the entire house. Overall, the structure has shown excellent thermal efficiency.

Another energy-saving feature is the solar-heated domestic hot-water system installed by Creative Energy of San Rafael, Calif. The solar-heating panel for this system is mounted on a south-facing section of the main roof, and caused some tricky plumbing runs. Electrical wiring was also difficult because so much had to be hidden behind the exposed structural members. Many of the wall sconces and pendant lights (photos above) were custom-made by artist Arthur Stern. This fine display of leaded and stained glass is very much in keeping with the Craftsman style that gives the house its warmth and character. □

Photos of completed deck: Ernest Braun, Courtesy of California Redwood Association

It's easy to think of a deck as a less than challenging project. After all, so many decks are simply rectangular platforms for a barbecue grill and lawn chairs. It's easy to think that decks don't present the technical challenges found in framing a new house or the exacting detail work demanded by a custom interior. But these attitudes all depend on your idea of a deck.

When remodeling contractors Fritz Hoddick, Mark Berry and Peter Malakoff agreed to build a redwood deck for Doug and Beth Rosestone, they set aside all the usual notions about decks and decided to build one that would use the skills and the approach to design that they apply to remodeling projects and cabinetwork.

Hoddick, Berry and Malakoff were in the midst of remodeling the interior of the Rosestone house (see "Craftsman Remodel," pp. 150-155) when the project was first considered. Handpicked mahogany, padauk, ebony and Douglas fir set the tone of their Craftsman-style interior work. The deck would have to complement those interior projects, and at the same time, it would have to withstand the weather.

Planning and design—The Rosestone house was built on a steep hill and had an excellent view of the valley below, but there was virtually no usable yard. So the deck would be what the yard was not: a place for outdoor entertaining, a place for the children to play and a place to enjoy the commanding view.

Eli Sutton, who designed the Rosestone interiors, called for a 2,138-sq. ft. deck with a freeform contour (drawing, facing page). Because several large trees poke up through the deck, it had to be constructed so that the decking could be cut back as the trees grow, without disturbing the framing.

A built-in bench, planters and a firepit/sitting area are among the amenities that add character to the structure, making it something more than a typical deck. The firepit/sitting area with its trellis echoes the Greene and Greene style found elsewhere in the building.

The deck had to be 4 ft. off the ground at one end and 17 ft. off the ground at the other. For safety's sake, Hoddick and his partners decided to over-engineer the deck supports using heavy timbers and liberal bracing. They used 1½-in. thick decking over joists spaced 2 ft. o. c. to reduce spring in the deck surface and provide a sense of solidity. One of the most common

A Rambling Deck

Laminated rails and Greene and Greene styling mark this redwood deck

by Christopher Grover

From *Fine Homebuilding* magazine (April 1987) 39:52-55

treatments used by Greene and Greene was the rounding over of edges and corners. And so on the Rosestone deck, all edges were radiused.

Construction—The deck stands on a grid of 6x6 posts set approximately 8 ft. o. c. To anchor the posts and distribute the weight of the deck, the builders poured concrete pads 2 ft. to 3 ft. below grade. These pads act as footings for the concrete piers that were formed with Sonotubes on top of them.

There was an existing 12-ft. by 12-ft. deck attached to the house. The redwood posts, ledgers and joists were still sound and serviceable so the builders decided to use these as a starting point. They also added new 6x6 posts, both redwood and pressure treated, to support all the additional square footage of deck. Since the redwood posts were more expensive, they were used only in the more visible areas. The posts were connected to the concrete piers with galvanized post bases (available from Simpson Strong-Tie Company, Inc., 1450 Doolittle Drive, P.O. Box 1568, San Leandro, Calif. 94577) embedded in the concrete.

Pressure-treated 4x10 beams rest on the posts and carry the load of the joists and decking. They're held in place with post caps that slip over the top of the 6x6 posts. They have a pair of flanges on top that form a channel and nailing surface for the beams. Where the deck sits high off the ground, 4x4 diagonal braces were lag-bolted in place between the posts and beams to eliminate sway.

Along the house, the builders lag-bolted ledger strips into the framing to carry the weight of the deck. Since the existing 2x8 redwood ledger was in good shape, it was left in place and reused, and new pressure-treated ledgers were added to supplement it.

The deck surface is the most visible part of any deck. Here, the builders used 2x6 construction-heart (con heart) redwood, which is virtually all decay-resistant heartwood with some knots and blemishes, and some manufacturing imperfections. They sorted the boards carefully, rejecting any that were twisted, cupped or split, or that had loose knots.

To keep nail heads from showing, most of the decking was fastened with galvanized finish nails, two per joist. The nails were driven in at angles forming a V for extra holding power, and then set. Using finish nails for decking can be risky since they don't hold as well as headed nails, especially once they're set. But redwood cups very little, and this minimizes the risk of deck boards pulling up against the nails.

The deck boards were run long all the way around the edge. After they were all in place, the curved outline of the deck was laid out using a rope with a pencil tied on the end to swing the curves. The builders then cut off the deck boards with an orbital jigsaw. The blade was long enough to score the joists underneath. Later, the scored joists were trimmed from below to match the line of the decking.

The fascia, which covers the ends and edges of the decking and runs all the way around the perimeter, was laminated from three ¼-in. thick by 10-in. wide redwood boards (photo right). It

Construction photos: Michel Winand

Rosestone deck floor plan

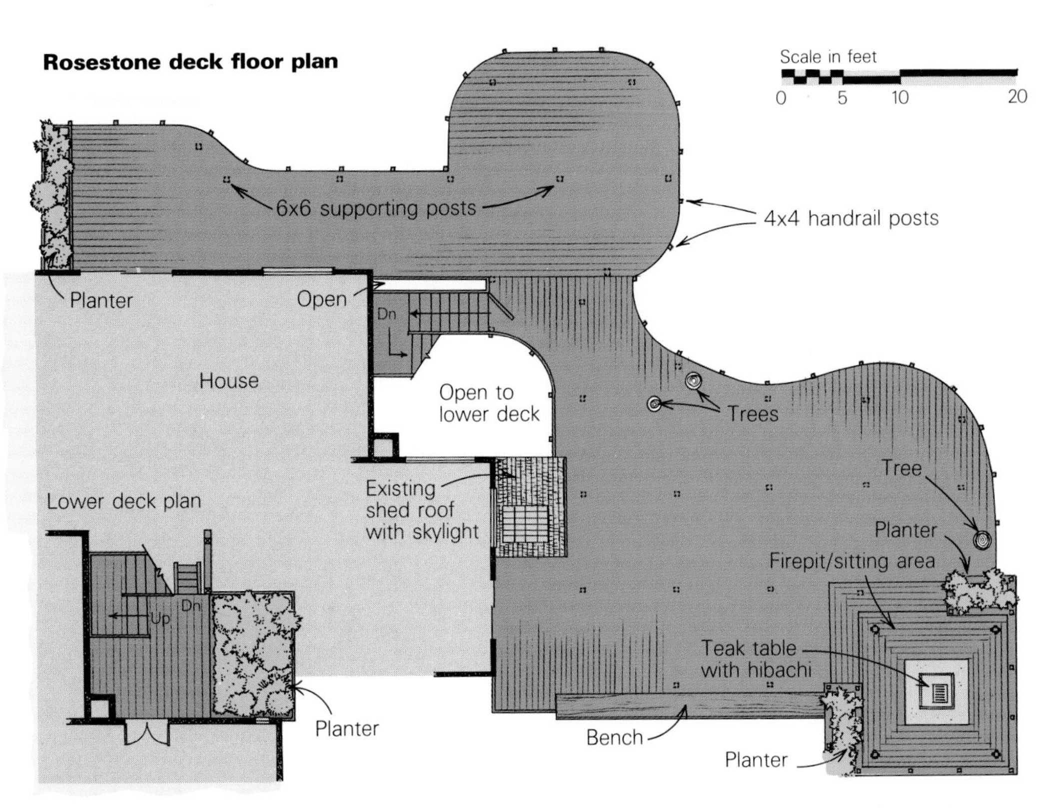

After cutting the curved outline of the deck with a jigsaw, the builders used marine-grade glue and staple guns to laminate three ¼-in. layers of redwood for the fascia around its perimeter.

was applied with a pneumatic nailer and Wilhold marine-grade plastic-resin glue (DAP, Inc., 855 N. Third St., Tipp City, Ohio 45371).

Railing—The organic shape of the deck's cantilevered borders demanded a freeflowing handrail to match (photo top left). The answer was to laminate the handrail from several thin pieces of redwood—an ideal wood for this because it bonds extremely well and comes in extra-long lengths, up to 28 ft.

The rail's 36-in. height was specified by code, the curves were determined by the shape of the deck, and the 5¼-in. o. c. spacing of the balusters was established to keep the family's small children from wiggling through.

The first step in building the rail was to lay out the 4x4 redwood posts, which were spaced 4 ft. o. c. and bolted to the fascia with carriage bolts. As the wood ages and shrinks the nuts can be tightened to take up the slack. Instead of using shims, the posts were plumbed by hand-planing the surface where they attached to the rim joist.

Laminating the two horizontal rails was the most difficult and time-consuming job on the whole project (photo bottom left). Using ⅜-in. thicknesses of redwood, the builders laminated an upper and lower rail. The upper rail is four plies thick, and the lower rail is three plies.

The first ply of the curved rail was attached to the 4x4 posts with nickel-plated screws and the same plasic-resin glue used on the fascia.

The second ply was attached using glue and pneumatic staple guns, shooting several staples every 3 in. to 4 in. along the length of the rail. The builders concentrated on getting the adjacent edges of the top and bottom rails flush because it would be difficult to use a plane in the small space between the rails. While gluing the rails, they used more than 90 Jorgensen pony clamps. After the glue cured (seven or eight hours), the next ply was applied.

Rails, balusters and bench. **Gluing up the rails required a lot of clamps, 90 in all (left). After the laminating was done, the edges of the rails were eased with a router. There are 380 balusters in the railing, and all of them were rounded over on a router table before they were installed (above). The redwood bench (below) runs for 23 ft. along one side of the deck. It is canted back slightly, at a casual angle that invites people to sit down, and on sitting, encourages them to stay.**

Subsequent layers were added in the same manner except that the outer layer was not stapled. The resistance of the wood to bending was strong enough to pull some of the posts out of alignment. On a few of the sections, a come-along (attached to a tree) was used to hold the posts plumb while the glue was curing. After all of the plies were glued up, the edges of the handrail were eased with a router.

It took two men six hours to dimension and cut the 380 balusters that surround the deck. They used a router table to radius the edges. Countersunk screws were used to attach the balusters to the rails and deck. Even the edges of the countersunk holes were rounded over by dropping a small roundover bit into them.

The rail and deck were finished with multiple coats of Penofin (Performance Coatings Inc., P.O. Box 478, 360 Lake Mendocino, Ukiah, Calif. 95482), a water-repellent penetrating oil finish that also contains a mildewcide and a little pigment to achieve a lasting "redwood" color. In the northern California climate, refinishing will probably be required in about two years.

Since the laminations are exposed to the weather, the builders treated the top of the handrail with paraffin to repel water. Using a small electric hot plate at the site, they melted the wax and brushed it on the handrail. As one person was doing this, another followed behind with a heat gun and reheated the wax to allow further penetration into the wood.

After the rail had weathered for several months, the builders checked carefully for any signs of delamination. On the few problem spots that were discovered, they used epoxy and clamps to bond the layers back together and then retreated them with paraffin. It has now been two-and-a-half years since the deck was completed, and the rail is holding up fine.

Bench—When seen from the deck, the 23-ft. bench appears to be floating in the air. It's actually supported by 4x4s centered under the seat every 4 ft., with horizontal crosspieces bolted into the top that cantilever in front and back to hold the seat (bottom right photo, facing page). As it runs along one edge of the deck, the bench not only provides plenty of seating, but also it eliminates the need for a rail in this area.

Firepit/sitting area—The firepit/sitting area is the most dramatic feature of the entire deck (photo bottom right). It is at the end of the deck closest to the ground and sits on an exposed-aggregate slab that was originally poured for a hot tub. At the center of the firepit is a concrete pedestal that holds a 34-in. square teak table with a built-in hibachi. There is also a teak box to cover the hibachi when it's not in use. The decking starts about 16 in. back from the edge of the table, at which point it is 16 in. off the concrete slab and makes a very comfortable sitting area around the table.

The sitting area steps up two times more (7 in. each time) before arriving at the level of the main deck. The changing levels help to differentiate this space from the rest of the deck, and the wide steps also make a nice place for potted plants.

The decking around the firepit was screwed and plugged instead of just nailed because it would be subject to the closest scrutiny. Multiple coats of Penofin were applied immediately after the decking in the sitting areas was installed. Months later, the wood still had a fresh redwood color.

The builders themselves weren't aware of how massive the trellis structure designed by Eli Sutton was until they were actually constructing it. The suspended screen alone weighed about 300 lb. when it was finished (photo top right). The trellis is supported by four redwood 6x6 posts, with radiused redwood 2x4s applied on all four sides of each post. The trellis itself is a loose gridwork of stacked, crisscrossing 4x4s and 6x6s, with all edges radiused ¾ in.

The 9-ft. square screen is suspended on 4x4s below the trellis. It's made of kiln-dried redwood 3x3s, put together with half-lap joints, glued and screwed from above.

With its several trees, changes in level and accommodations for planting, cooking, playing and just sitting around, this deck feels more like a backyard than a wooden platform. □

Writer Christopher Grover makes his home in Fairfax, Calif.

Even though it was made with kiln-dried redwood, the screen suspended below the trellis (above) weighed nearly 300 lb. and had to be hauled into place with a come-along attached to cribbing stacked on top of the trellis. The firepit/sitting area (below) is centered around a teak table with a hibachi set into it. The lower levels, solid walls on two sides, and trellis overhead make a private and cozy space for outdoor dining.

Above, Wallen II is located high on a windswept hill overlooking San Francisco Bay. Its exterior materials include cast-in-place concrete walls, corrugated steel roofing and steel factory windows. Bold massing of shapes makes the house seem larger than 1,300 sq. ft. Photo by Jane Lidz.

Bernard Maybeck's Wallen II House

This 1937 house designed by a California architect can still inspire those struggling to achieve quality with economy

by Thomas Gordon Smith

Bernard Maybeck was an architect of incredible inventiveness, known throughout his long career for using a variety of forms and materials in unconventional ways. When he designed Wallen II, an inexpensive house built by its owners, Maybeck turned to industrial materials and construction techniques.

Two of the ten houses Maybeck designed during the 1930s were built for his son Wallen and Wallen's wife Jacomena. The first house was built within the family compound in Berkeley, Calif. After living there two years, the younger Maybecks decided they wanted some privacy and moved away. Maybeck and his wife then moved into Wallen's house.

In 1937 Wallen and Jacomena had to leave their retreat, and they returned to Berkeley hoping to move back in. The elder Maybecks wouldn't budge, and Jacomena recalls her father-in-law saying, "You're wet and we're dry; we'll give you an acre and you build a house."

The house had to be economical; a nucleus had to be quickly constructed for occupancy, and it had to be built of fire-resistant materials, as the dry grass on the site was a fire hazard.

The plan called for three elements: two pavilions connected by a wooden portico, and a detached garage (shown on the opposite page). Together they would form a courtyard enclosed by a brick wall. A scheme was devised by which the west wing, containing one bedroom and a garage, would be inhabited first—the bedroom for sleeping and the garage (temporarily) for both eating and sleeping. In the second phase, a studio wing would be built perpendicular to the first. It would contain a living room, a dining-

From *Fine Homebuilding* magazine (April 1981) 2:18-25

Above, the north end of the bedroom wing is pierced with glass block. The diamond-shaped voids were cast into the concrete wall with special steel and wood forms. Ivy-covered walls, the steep gable and the mullioned windows give the house a look of a 17th-century country manor. Photo by Jane Lidz.

Standard factory sash, painted red, were used as windows. Especially beautiful is this bay at the west end of the living room.

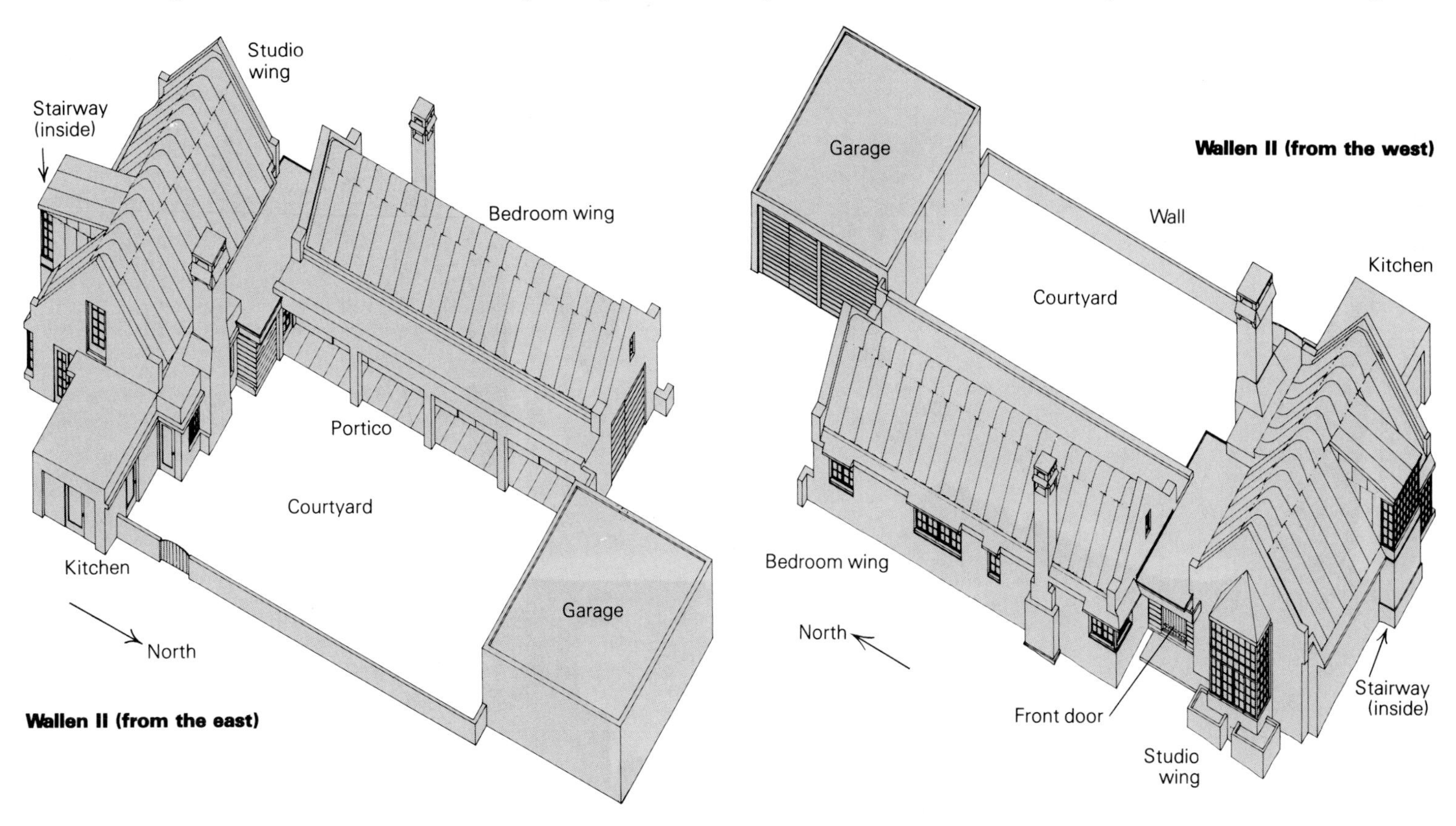

inglenook area and a kitchen. Finally, the main bedroom wing was to have been constructed to form the courtyard. The first two wings were built as planned, but the third never was. Still, the house is true to Maybeck's idea: separate elements huddling together for protection on an exposed site in a fierce and windy climate.

The site—Wallen and Jacomena chose a site that Maybeck liked: it is on the ridge that divides the cities of the eastern San Francisco Bay area from the wild inland canyons. The site, like May-

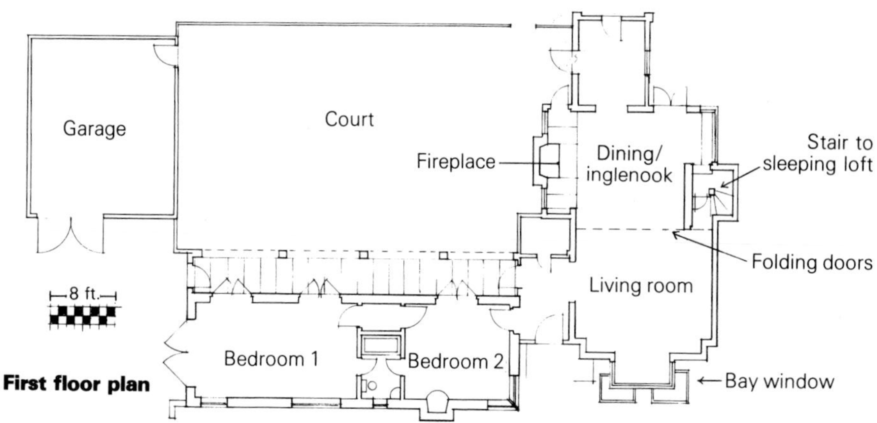

beck's temperament, is balanced between the urban and the rustic. Even today the site seems isolated and conveys the feeling of a windswept Yorkshire heath.

Maybeck knew that the weather was more severe on the ridge than lower in the Berkeley hills, and he emphasized the need for a low-lying, well-insulated house built of rugged materials to resist the west wind and the strong easterly gusts that blow in the fall, and the hot, sunny skies. However, the house is not merely an example of form following climate. There is a sensitive mesh between the image of the house and its windblown site. Maybeck must have had an isolated manor house in mind when he began his design. It evidently conjured up the same image to the locals who called it Wuthering Heights.

The design—The materials and methods of construction specified for Wallen II could have lent the house a forbidding air. Maybeck overcame this obstacle by deftly handling the house's various forms, and by giving priority to a warm, well-scaled interior.

Concrete walls were cast in place using a system of modular metal forms devised for industrial construction. Although the house has only 1,300 sq. ft. of living space, it appears to be larger—a result of its monumental forms and rugged material.

Corrugated steel roofing is set between the projecting concrete gables. Stains from the roof have discolored concrete projections that protect the windows. The construction drawings show that concrete tiles were intended to sheath the roof for additional fire protection. These were omitted, however, because they cost too much, and because Maybeck liked the look of corrugated steel.

Steel windows are set in flat wall planes. Their thin mullions are like delicate drawings on the massive surface. But this is an owner-built house, and the dullness of industrial fixtures is avoided by creative detailing. Although the house is solidly built, the texture of human error is everywhere, from conception in the preparatory sketches to the improvization of details during construction. The solutions are often too casual, but even problem areas are usually saved by the strength of the idea. For example, a recessed ceiling light fixture is turned upside down and positioned atop the newel post at the stairway

Above left, the stair leads to a sleeping loft. The subtle color difference between the concrete beams and the bubblestone panels can be seen on the wall to the right. At the west end of the living room (photo at left) the bay window provides a spectacular view of San Francisco Bay. Sliding glass doors shut off the bay in cool weather. At right, the living room merges with the low-ceilinged inglenook. The folding doors used to isolate the inglenook in the cold winter months are stacked against the wall. The joint lines of the insulating bubblestone were left on the walls, and natural finishes are used throughout.

Casting walls for west wing

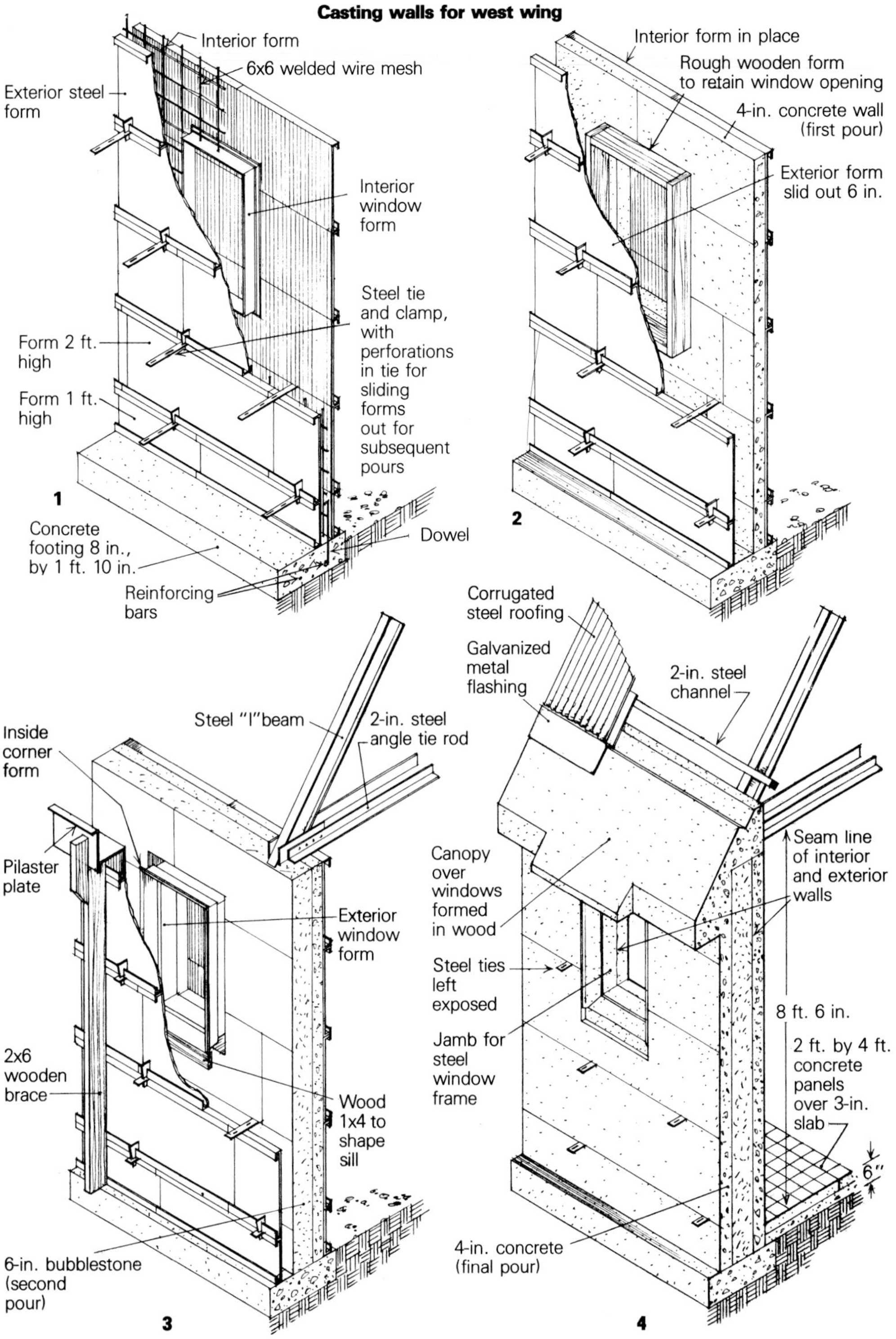

1

Interior form
6x6 welded wire mesh
Exterior steel form
Interior window form
Steel tie and clamp, with perforations in tie for sliding forms out for subsequent pours
Form 2 ft. high
Form 1 ft. high
Concrete footing 8 in., by 1 ft. 10 in.
Dowel
Reinforcing bars

2

Interior form in place
Rough wooden form to retain window opening
4-in. concrete wall (first pour)
Exterior form slid out 6 in.

3

Inside corner form
Steel "I"beam
2-in. steel angle tie rod
Pilaster plate
Exterior window form
2x6 wooden brace
Wood 1x4 to shape sill
6-in. bubblestone (second pour)

4

Corrugated steel roofing
Galvanized metal flashing
2-in. steel channel
Canopy over windows formed in wood
Seam line of interior and exterior walls
Steel ties left exposed
Jamb for steel window frame
8 ft. 6 in.
2 ft. by 4 ft. concrete panels over 3-in. slab
6"
4-in. concrete (final pour)

Maybeck frequently found beauty in inelegant materials and simple methods. For the front door Maybeck laminated 1x6 redwood planks to a sheet of plywood, and banded it with steel straps torched from flat stock (left). The edges were not finished. Inexpensive strap hinges on each band support the door.

(photo page 81). The construction is spontaneous and crude, but the idea is wonderful.

The 900-sq. ft. studio took almost a year to build. It was more complex than the west wing, and required more finish work. It consists of a living room with a high, trussed ceiling, a sleeping loft and a low-ceilinged dining area. A deep inglenook, essentially a room-sized hearth into which chairs can be drawn, is located at the north side of the dining area. A large window to the south floods the area with light on sunny days. (The kitchen, a flat-roofed block tacked onto the rear, has been engulfed in subsequent additions.) Maybeck wanted to make the rooms of the studio wing seem enclosed and secure against the weather. He provided a large bay window to the west to take advantage of the spectacular panorama of San Francisco, the Golden Gate, and Mount Tamalpais. From the exterior, this bay set in the soaring facade is a focal point of the composition.

The construction—Maybeck met three concerns head-on in this house: economy, energy efficiency and flexibility. The Wallen Maybecks were guinea pigs for a number of technical innovations that must frequently have been rationalized on the basis of economy or fire resistance. The principal experiment was to employ a system of modular forms for concrete construction that had been developed for industrial applications by a friend of Maybeck's, A.E. Troeil. The Steel Speedform Method consisted of 2-ft. high flanged steel panels of various widths. The panels were clamped together with steel hooks. These were inserted into regularly-spaced perforations in the steel ties that spaced the forms (drawing at left). The Troeil forms were designed for casting several plies of concrete, poured in successive operations. Maybeck designed the 14-in. thick walls of the west wing to be composed of three layers; the interior form was left in place after the first pour had cured, and the exterior forms were pulled out along the ties once for each subsequent pour. The sequence of pouring these walls is shown at left.

Special modular shapes existed to form 90° projections and to turn 90° corners. Troeil had also developed steel forms that were placed between the continuous steel panels to create door and window openings. When the concrete was cured and the forms were removed, the door forms left a steel doorstop embedded in the concrete, and the impression of a curved molding around the jamb and head. Maybeck employed Troeil's module when it was convenient, and built wood forms to meet other conditions.

The sequence of setting up forms had to be efficiently planned. After the footings had cured, the interior steel panels were set in place to a height of 9 ft. The interior window forms were positioned, conduit and switch boxes were routed for electrical service within the walls, and 6-in. by 6-in. wire mesh reinforcement was

Maybeck's fireplace (far right) is truly no more than an indentation in the dining room's concrete wall, with the fire on a hearth within the room. The chimney (right) begins as an open hole in the ceiling. It is capped with a special form that Maybeck devised for keeping out rain. The shape creates a venturi effect, increasing the vertical draft enough to prevent rain from falling into the chimney. This works well with a raging fire, but not too well in a summer downpour. Both photos, at right, by Jane Lidz.

put in place. The exterior forms were set up and connected by steel hooks at the ties. Then the 4-in. thick wall was poured. When the concrete had cured, the exterior forms were slid out 6 in. along the ties and fastened again by the hooks. The exposed face of the concrete wall was covered with bituminous waterproofing and the forms were ready for the second pour.

Maybeck filled the second volume with "bubblestone" mixed with rice hulls. Bubblestone was an air-entrained insulating cement that had been developed by one of Maybeck's friends, John Rice. (Evidently rice hulls could be purchased cheaply, and they were added to the bubblestone to increase its insulating value.) There is no indication in the drawings that any reinforcing was provided for this layer of material. Presumably, rough wooden forms were built around the perimeter of the interior window openings. The forms provided adequate space for inserting the exterior window forms and accommodating the finish concrete.

When the insulating cement had cured, the exterior forms were slid out another 4 in. and positioned for the final pour. This operation must have been more exacting than the previous two, and the steel and wooden forms had to be set up in special ways. For example, a projecting canopy was provided for at the top of the third wall by substituting standard 8-in. corner panels for the top course of 2-ft. high flat panels. The additional projection of the canopy over the windows must have been done in wood. (See drawing on the opposite page, step 4.) The edge of the canopy meets a vertical fascia over which the concrete slopes back at a 51° angle for a depth of 1 ft. 2 in. Thus the third pour formed a triangular cap, which bonds the three thicknesses of the wall together. The parapet gables were formed for their full thickness during this pour. Both the gables and the rake cap were poured to encase the ends of steel I beams and channel purlins that were set on the first and second plies prior to the third pour. Once the third pour had cured and the forms were stripped, little finish work had to be done before the west wing was habitable. The corrugated roof had to be applied, the ceilings hung and insulated, doors hung and windows installed.

The walls of the garage were cast by a less laborious process. Troeil had devised a system for building with concrete and glass block, shown in the drawing at right. Panels of 2x4s set in a diagonal lattice were the key. These forms

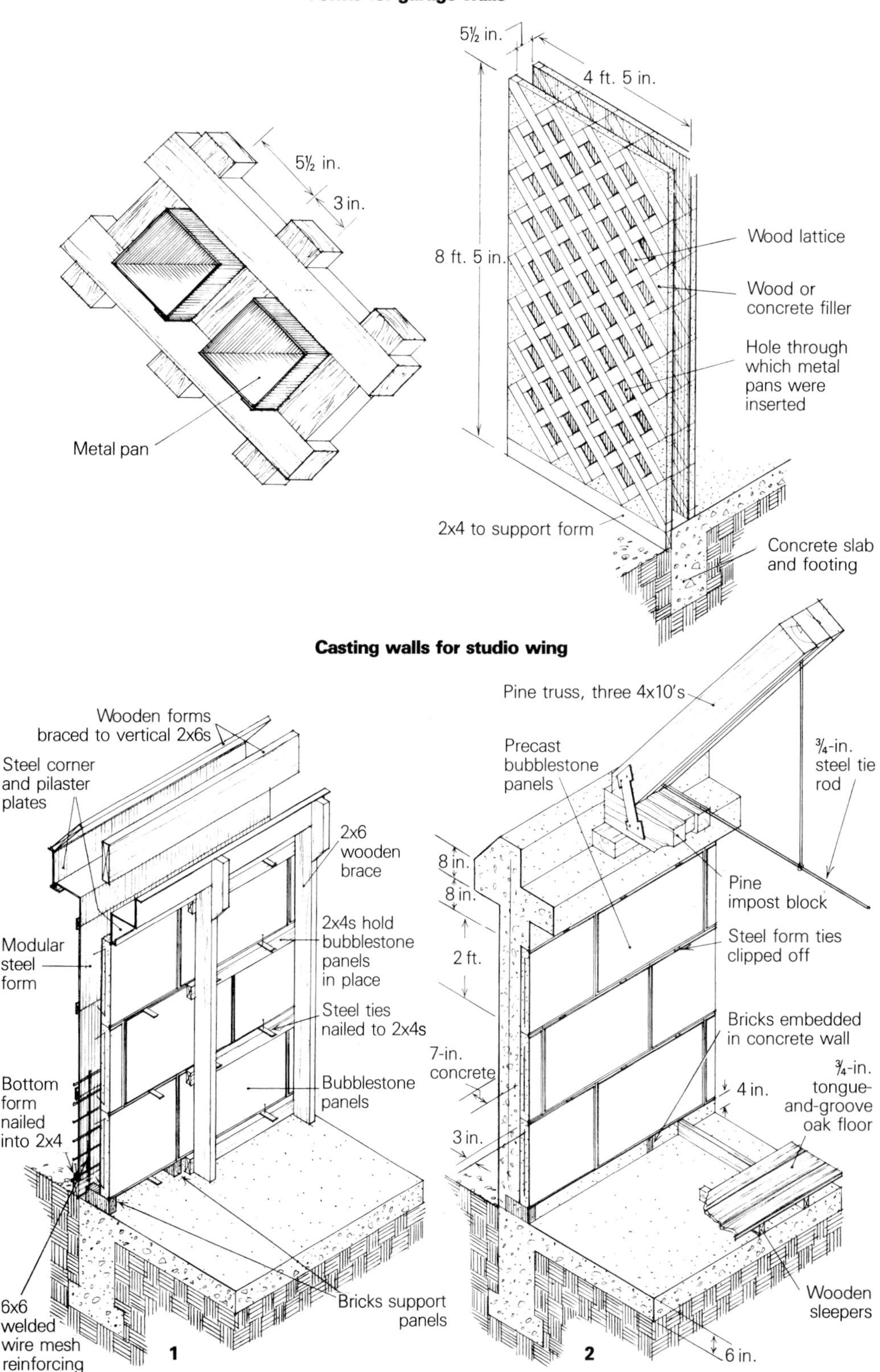

Bernard Maybeck

Like Wallen II, Maybeck was mysterious and romantic. He's best known for designing the San Francisco Palace of the Fine Arts in 1915. Although he was born in New York City and studied in Paris at the Ecole des Beaux-Arts, it was in the California sun that he matured as an architect.

Educated in the sophisticated French tradition, he learned to design with elements of historical architecture in an eclectic, creative way. Maybeck was especially taken with the Beaux-Arts' love of color in ancient and medieval architecture, and with classical Greek and Roman architectural forms. He traveled during his student years and sketched simple vernacular buildings as well as Romanesque and Gothic churches.

When he left Paris in 1886, he spent the next three years apprenticing in different architectural offices before settling in the San Francisco Bay Area. There he developed an eccentric design style, combining his Beaux-Arts training with Japanese, Arts and Crafts, Chinese, Swiss, and any other architectural influence that interested him. He was a truly eclectic designer, thought of by many as a Bohemian mystic.

When Maybeck was in his forties, he produced a series of houses in a wide range of styles, from the Lawson house in 1907 (a Pompeiian villa with pastel and black-tinted stucco over concrete walls), to the Leon Roos house in 1908 (a version of a Tudor half-timbered building). Maybeck's inventiveness with structure and form, and independence from the conventional use of historical styles appear in all of his buildings. His use of motifs considered inconsistent with each other kept his critics constantly off-balance. For example, his Faculty Club on the University of California campus at Berkeley (1901) is a low, tile-roofed Mediterranean building on the outside, with a steeply-pitched wood Viking hall inside. He never feared taking a chance.

Through the 1920s Maybeck remained a successful architect, working on some of the largest projects of his career. In 1932, at the age of 70, he retired from his architectural practice. That is when the story of Wallen II begins. —*T.G.S.*

Folding doors (top left photo) separate the inglenook and living room. The center panel is hinged, and opens like a standard left-hand door. The doors are 24-in. plywood panels on one side, and finished to match the ceiling on the other. A shelf (center photo) was cast into the top of the wall to support the ceiling rafters in the studio wing. The built-up rafters are bracketed to blocks that would normally be the bottom chord of the roof truss. Maybeck cut the chords and shaped them to form decorative elements. A $^{3}/_{4}$-in. steel tie rod, painted red, forms the actual chord. Steel framing (left), designed for industrial buildings and barns, is exposed in the attic of the bedroom wing. Purlins are bolted to the steel rafters; sheets of corrugated roofing are, in turn, bolted to the purlins, and rubber washers on the exterior provide a waterproof seal. A curved section of corrugated steel caps the roof peak.

Maybeck was a very inventive detailer, and the house is filled with little surprises. At the top of the stair to the loft (above left) a ceiling light fixture recessed in the newel post was used as a lamp. A standard hot-air heater register is the vent. The newel post is a box made up of four 2x8s. Glass blocks are set into a concrete grid (above right) in the kitchen. The process of casting the grid is described on page 79. Color photo, above, by Jane Lidz.

were set up side by side for the required length, with a parallel row 5½ in. away. Metal pans were inserted in each opening to span from one form to the other. Wire reinforcing seems to have been inserted next, and the walls were cast by pouring concrete from the top. By gravity and tamping, the concrete made its way to the bottom of the form and gradually filled it up. Once the concrete was cured, the metal pans were knocked out and the wooden forms were stripped. This left a grille pattern with open holes. A similar method was used to cast the kitchen walls, where, after stripping, the holes were filled with glass block (shown above, at right).

The walls of the studio wing were cast using yet a third method. (See page 79.) The same modular steel forms were used for the exterior walls. However, the 10-in. thick wall was cast in one operation, and no steel panels seem to have been used on the interior. After the floor slab and footings had cured, bricks were placed every 18 in. to support precast 2-ft. by 4-ft. panels of the bubblestone and rice hull mixture. Three rows of bubblestone panels were laid up and supported by vertical 2x6s. The wooden members also supported steel corner and pilaster forms to cast an overhang similar to the canopy of the west wing. After electrical conduit and reinforcing had been installed, the exterior forms were positioned as in the west wing.

The studio wing required more finish work than the west wing. The roof was attached to built-up 4-in. by 10-in. wood beams supported by wooden blocks on the concrete wall. The wooden floor of the sleeping loft was suspended from the beams by vertical steel ties. On the main floor, tongue-and-groove 1-in. by 6-in. oak flooring was nailed to 2-in. by 3-in. sleepers attached to the slab, creating a small, under-floor air plenum. Nonetheless, much of the interior was finished as soon as the forms were stripped. Members of the family did what painting and cleaning they could to speed the project along. Once the studio wing was finished, it was linked to the west wing by a wooden portico, and the house was complete.

The economy of the Wallen II House was partially achieved by selecting materials that were inexpensive at the time, such as particle board, plywood and concrete. However, it was mainly done by switching standard priorities on various aspects of construction. The percentage of the budget spent on finish work was exceptionally low in relation to the cost of the structure. An invoice for the construction of the west wing (up to pouring the bubblestone wall) shows a cost of just over $1,000. This suggests that the process was quite labor intensive.

Mechanical systems—Maybeck was an advocate of hardy living, and he believed in underheated houses. His solutions in Wallen II follow this inclination toward simple and conservative methods of heating. Although a water heater supplied hot water to radiators throughout the house, the intention was to focus the warmth in one area: the low-ceilinged dining room and the north-facing inglenook in the studio wing. The family changed its activities with the winter cold and congregated near the open hearth. The wooden ceiling was insulated with redwood bark, and an insulated plywood partition could be pulled across the inglenook to contain the warm air. (See top photo on the opposite page.) The partition can be secured to the floor for rigidity, and two of the panels slide into a head and form the jambs for a rigid door frame.

Maybeck designed a remarkable Rube Goldberg heating system for the dining area. He routed the pipes from the hot water supply through a Ford Model T radiator that was installed under the stairway to the loft. Water circulated through this radiator and then went to the normal heating radiators throughout the house. A blower forced air through the Model T radiator into the plenum between the concrete slab and oak floor. This system kept the enclosed area warm and snug.

The Wallen II House is an unusual building, uncommon in material and style. It is remarkable that in the late 1930s, when the International movement in architecture was beginning to make itself felt throughout the United States, Bernard Maybeck designed a house in the 17th-century Jacobean style.

Yet, the house also reflected bold innovations in the use of materials and spatial arrangements that were popular with his younger contemporaries. Wallen II remains a modern building. Many of its innovations may seem unique to its site and climate; however, the house solved the perennial problem of building cheaply without sacrificing quality. □

Thomas Gordon Smith, winner of the Rome Prize, practices architecture in San Francisco, Calif. A visit to Wallen II as a teenager inspired him to become an architect.

Mullgardt Rescue

Debauched by careless remodeling, a Craftsman style house is brought back to its former glory

by William Dutcher

In 1971, Bob and Anita Stein purchased a Craftsman-style house in Piedmont, California, which had been designed by Louis Christian Mullgardt (see sidebar, p. 86). It was one of only two houses that Mullgardt conceived in this style, and it was built as a model home in 1907 (less than a year after the infamous earthquake that set San Francisco afire) by an enterprising contractor named C. M. "One Nail" McGregor, It was also Mullgardt's first-built house (photo on p. 86).

Bob grew up in the East Bay, where he had been exposed to the Craftsman style houses of such respected designers as Bernard Maybeck, Ernest Coxhead and John Hudson Thomas and had a latent interest in the Craftsman period and style. Once he and Anita had taken possession of the Mullgardt house, his interest intensified, and being a history buff, Bob set about piecing together a genealogy of the house.

In addition to compiling a library about architecture and design of the period, Bob and Anita searched out former owners of the house and put together an oral history. It recounted various changes and remodelings that had taken place over the years, and more important, what the house had been like in its original form. These verbal descriptions (by people not especially attentive to architecture or building) were supplemented by old family snapshots, which proved invaluable in discovering and recreating lost details.

Early on, Bob met Robert Judson Clark, an architectural historian at Princeton who had done research on Mullgardt. This led to a visit to Mullgardt's Evans house in nearby Mill Valley, which is very similar in design and detailing to the Piedmont house, and virtually unchanged. It was there that we could get the full flavor of what the Stein's house had been, and could be again. Original drawings of the Mill Valley house were found in the College of Environmental Design archives at the University of California. The drawings helped in determining the dimensions and proportions of trim details.

I came on the scene in 1974 when the Steins asked me to assist them in the restoration and renovation of their house. The first and most obvious task was to develop drawings of the house as it currently existed, and as it had been when originally built. While the former was relatively easy, the latter proved more difficult.

Restored to its original state, the northeast corner of the living room (photo right) exemplifies the level of finish and detail found throughout the house. An abstract dragonfly composed of painted squares and gilded linen paper is repeated in the frieze located above the plate rail in the living room. Burlap glued to wallboard is the finish in the bays between the box beams. Hung over the mantelpiece is a portrait of architect Louis Christian Mullgardt, who designed the house at the turn of the century.

Sleuthing—The house had been remodeled and redecorated more than once, with the indoor spaces suffering more than the exterior. The principal rooms—entry, living, dining, and stair gallery—had been considerably altered. The clinker-brick fireplace facing had been covered with bathroom tile painted black. Features such as plate rails, brackets and dentils had been removed, and the remaining redwood trim had been painted. The living-room window seats and bookcases had been torn out, windows closed up or replaced with doors, and the glazed partitions between the living room and entry had been removed. The hardwood floors in all rooms had been covered with a repulsive green shag carpet.

Virtually nothing remained of the original kitchen/pantry area, which now included orange plastic counters, aluminum windows and sliding patio doors from a '50s remodeling. An adjacent "servants'" room with fireplace still existed. We were unable to uncover much information about these areas, other than their overall form, and I assume they were originally finished to a minimal level as was customary at the time.

The old snapshots indicated many materials and details no longer in evidence, but suggested that some might still remain, hidden from view behind fields of drywall. Knowing that the architectural trappings of another era might lay just under the surface infused the project with the mystery of an archaeological dig. Because Bob and Anita would be living in the house while work progressed, I started my "excavations" tentatively so as not to cause too much disruption. By removing trim pieces such as switch plates and heating registers, I discovered that some paneling remained beneath the new drywall. Emboldened by this discovery, I pried away patches of drywall in various locations and found the battens had been removed but the boards remained, and the painted trim was original. Removing drywall from the living-room ceiling revealed the original burlap fabric between the beams, which had covered 1x6 T&G redwood ceiling boards.

With the drywall removed, we could detect ghostly outlines of brackets, plate rails and other features that had been applied over the paneling. The exposed wood had darkened more than the covered areas, leaving faint but discernible images that could be measured. All wall and ceiling surfaces had been redwood boards, either exposed board-and-batten paneling as in the living room, or boards butted together and covered with painted canvas as in the bedrooms and bathrooms.

By combining the information that we learned from measuring the outlines of former trim pieces with the precise drawings of the Evans house in Mill Valley, I was able to produce working drawings that closely approximated the house's original details. Pan-

From *Fine Homebuilding* magazine (Spring 1989) 52:78-82

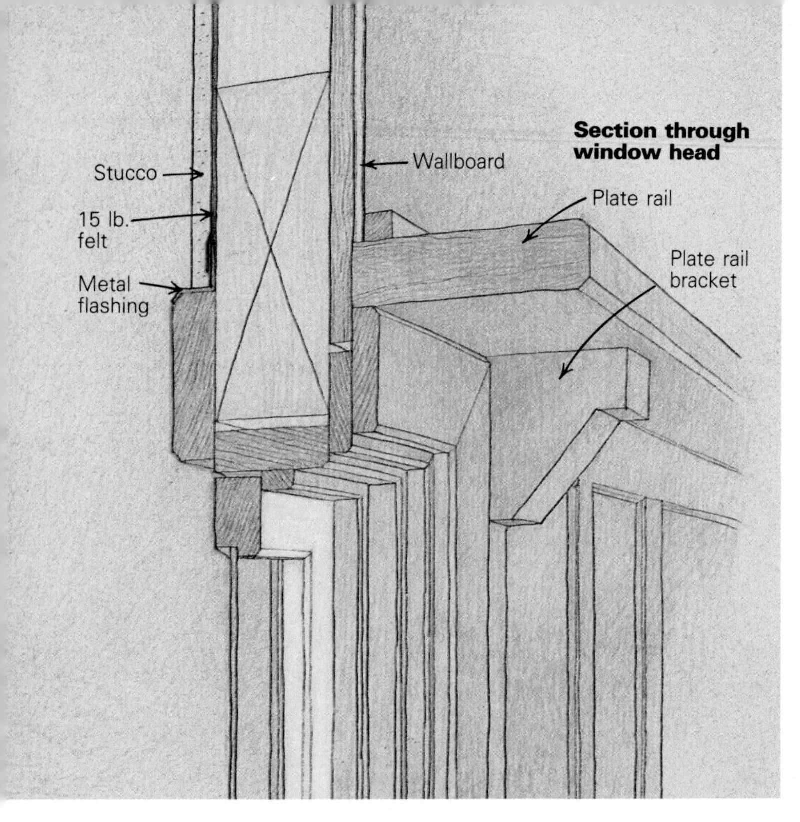

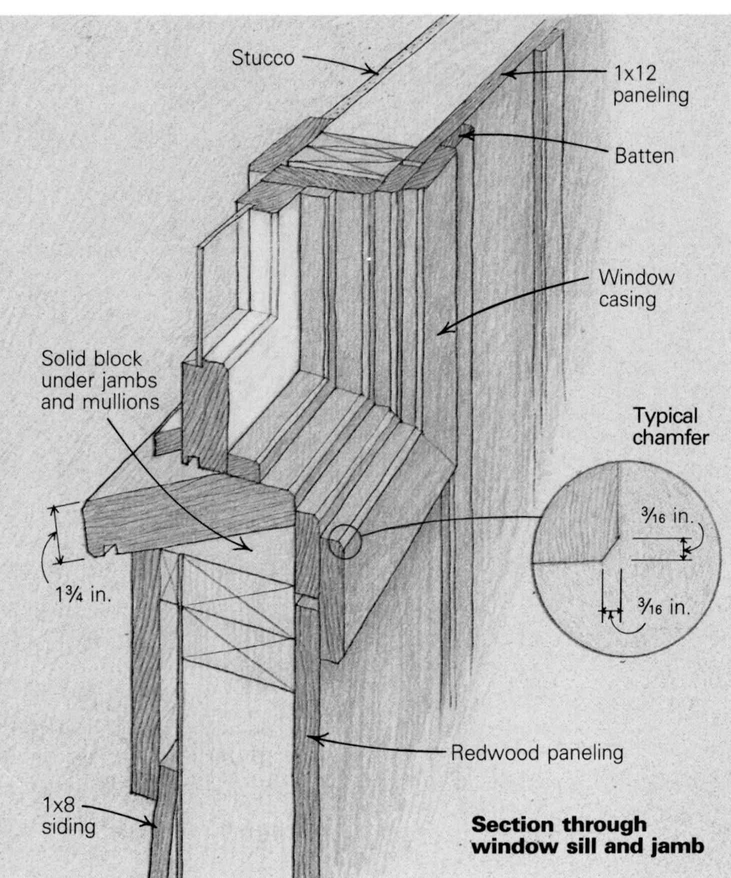

A two-sided fireplace makes a partial screen between the eating area and the sitting area. Its hood is made of planished copper, a favorite material of Craftsman-era metalworkers.

eling, interior, and exterior trim are redwood, and the individual trim members are chamfered and offset to create gangs of parallel shadows and highlights. A section through the window on the west wall in the living room shows how these hierarchical trim pieces are typically assembled throughout the house (drawing above).

Resurrecting the living room—Virtually all the carpentry work in the house was done by Arne Fremmersvik, a carpenter who began his career in Norway over 60 years ago. But before Arne could begin, the painted redwood had to be brought back to life. Fortunately, most of the wide paneling boards had been shielded from the brush by drywall. The trim pieces and the ceiling box beams weren't so lucky. Bob hired a number of people to apply paint remover, scrape it away and then sand the redwood with orbital sanders where they could reach, and by hand where they couldn't.

Once the original woodwork had been resuscitated, Arne started filling in the pieces that had been torn out. In the living room, he put back the window seats at the north end, and the bookcases that flank them, making a secluded inglenook by the fireplace (photo previous page). The new pieces of redwood trim were decidedly lighter in color than the original boards, so a wax finish was applied that helped to even out the color differences. All wood surfaces in the house, except for those in the kitchen, were finished with two coats of Johnson paste floor wax. We used the kind that has a terra-cotta red pigment in

Elevation looking north

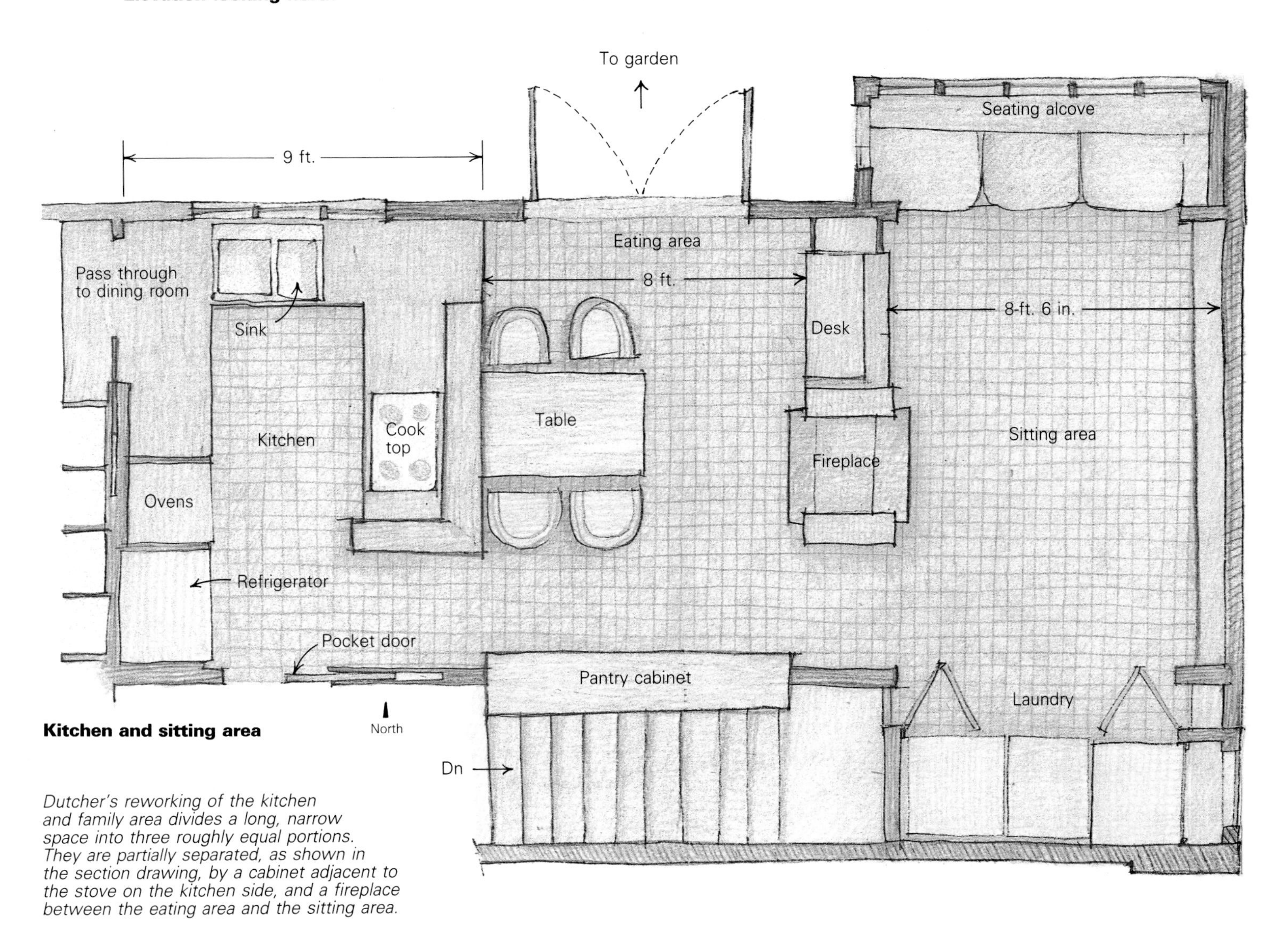

Kitchen and sitting area

Dutcher's reworking of the kitchen and family area divides a long, narrow space into three roughly equal portions. They are partially separated, as shown in the section drawing, by a cabinet adjacent to the stove on the kitchen side, and a fireplace between the eating area and the sitting area.

it and applied it, unthinned, with rags. On an earlier job, I learned not to use the unpigmented wax on redwood because it leaves a white residue that comes off only with mineral spirits.

After the first coat of wax had been applied, the crew filled nail holes with pigmented putty. After the final coat of wax, it was still possible to see differences in color between the new and old wood, but time has been the cure. As they have taken the light together, the woods have evened to the same tone.

Sconce and frieze—As might be expected, the prior remodelings had seen the demise of all the original lighting and plumbing fixtures, as well as hardware. For reasons that are mysterious to me, the Craftsman style has never been as popular as other styles, such as Victorian, and consequently there are few reproductions to speak of. Salvaged materials are equally hard to come by. We found that salvaged light fixtures, when available, were equal in cost to making new ones. So we took the opportunity to search out and employ talented artisans to build new fixtures in the Craftsman style.

In the living room, new sconces are mounted above bookcases (photo on p. 82). We had originally thought to have them made out of metal, but Bob preferred to see something in a contrasting wood. So I designed the sconces after similar fixtures designed by the Greene brothers, and furnituremaker Steph Zlott made them out of Peruvian walnut.

The frieze that surrounds the living room between the plate rail and ceiling was origi-

Louis Christian Mullgardt

Louis Christian Mullgardt (1866-1942) practiced in the San Francisco Bay Area from 1905 to the early '20s and played a major role in the development of what came to be called the Bay Area Tradition of architecture. His earliest houses had wood frames, with exteriors composed of stucco, clapboards and half-timbering. During the next five years he continued to work with these materials, refining the look of the houses with roughly trowled stucco exteriors and heavily bracketed shallow roofs. These buildings reminded some observers of California Mission architecture accented with Tibetan details.

Mullgardt's most famous residence was the Henry W. Taylor house—an enormous stuccoed fortress surrounded by tiers of roofs and retaining walls atop a conspicuous Berkeley hillside. But its monumental scale wasn't enough to get it past the depression years: it was torn down in 1935. In the words of the late architectural historian, John Beach, Mullgardt's built work "seems to exert an irresistable fascination upon bulldozers."

Mullgardt's fame attained its peak in 1915 at the Panama-Pacific International Exposition, where his "Court of the Ages" depicted the evolution of mankind in plaster details, columns and an altar-like tower. Soon after, he designed several important Bay Area buildings, such as the president's house at Stanford University and the M. H. deYoung Memorial Museum in Golden Gate Park.

After World War I, Mullgardt fell on hard times. His romantic parables in plaster were completely out of sync with Modern Architecture, and his turn to spiritualism after the death of a son further eroded his reputation. He died in 1942, in a state hospital in Stockton, California.
—Robert Stein

The first house built by the Bay Area architect Louis Christian Mullgardt stands on a hill in Piedmont, California. Its battered wall base is typical of Mullgardt's residential designs, as is its light band of stucco over the dark wood. Mullgardt was fond of squares, and the patterns displayed in the windows were repeated in various motifs throughout the house.

nally covered with gilded burlap (as were the bays between the box beams). But rather than repeat this color scheme, I saw an opportunity for a more decorative treatment, while at the same time introducing some color into an otherwise monochromatic environment. Throughout the house, Mullgardt had used repeating patterns of squares like those of the window panes and the ventilation grills. Picking up on this theme, I came up with a stencil pattern that abstracts in squares a dragonfly form. Painter Bob Laney did the friezework, using a stencil as a layout guide. Then he filled in the colors with a brush. The gold-leaf portions of the pattern are squares of linen paper, gilded first and then glued to the painted drywall.

To get the right blue for the frieze background, I bought several cans of paint and started mixing. I applied likely tints to 20-in. by 30-in. pieces of illustration board and then looked at them in the Stein's living room, right on the wall. In my experience, it is crucial to look at colors in the environment in which they will be finally viewed. I gave the final color to Laney, and he had it duplicated by a paint store that uses a computer matching system to analyze color samples.

Tackling the kitchen—Of all the areas in the house that needed attention, the kitchen and its adjacent spaces presented the most challenging problems. At one point the kitchen consisted of a string of three small rooms: one for food preparation; one for a pantry; and a third for servants' quarters. While we could have determined exactly where the walls had been, we didn't really care to know because the original plan had nothing to do with the way the Steins intended to live in the house. In general, kitchens and similar "utility" spaces were given considerably less importance around the turn of the century than they are today. We gutted the kitchen area, and turned the resulting space into one large "family" room that has three separate areas: one each for cooking, eating and sitting (drawing previous page).

Each space is compact—about 100 sq. ft. apiece. But we kept them open to one another by defining their boundaries without using walls. The kitchen is separated from the eating area by a 54-in. high cabinet behind the cooktop. A table abuts the cabinet, and because the high cabinet screens the kitchen counters from sight, diners can concentrate on the view to the garden through the French doors or on the fireplace that separates the eating area from the sitting area.

The original house had a fireplace in this same location, but it was tucked in the servants' room and hidden behind the kitchen wall. We decided to take advantage of the situation and make the focus of the room a new two-sided fireplace (photo p. 84). It was built of sand-molded brick and includes a planished copper hood made by silversmith and sculptor Louis Mueller. We added a small projection for a window seat in the sitting area.

The kichen area got all new windows of the same size and detail as others in the house, and the walls were paneled in the same board and batten used in the principal rooms. The ceiling was reconstructed in the same manner as the adjacent dining room. The bays between exposed beams, however, were left as natural wood rather than covered with burlap, making the surface easier to maintain and creating a more informal presence. Unlike the other woodwork in the house, the kitchen got two coats of McClosky's Heirloom flat finish before receiving the two coats of wax.

The completed house stands as a reminder of a time when abundant natural materials and craftsmanship were readily at hand for those who loved them. It is sad that we are rapidly losing the craft traditions in the building industry. In searching out craftsmen, I found some pursuing careers in the fine arts, rather than in the building trades. Even when they worked with unfamiliar materials, and usually only for the income it would bring, they approached the work with the same intensity and skill with which they made their art, resulting in craft of the most special and beautiful kind. □

Architect William Dutcher is located in Berkeley, California.

Resurrecting the Bolton House

There's more to restoring a Greene and Greene house than following the original plan

by Ken Ross

Black-and-white photos: Courtesy College of Environmental Design, Documents Collection, U.C. Berkeley

When I moved to Pasadena, Calif., in 1974, I saw for the first time several homes designed by Charles and Henry Greene. It was love at first sight. So when one of their houses came on the market in 1979, I bought it without so much as a backward glance.

To my eye, the Greene brothers' work is the ultimate expression of the American Craftsman movement in architecture. Their work brings together several design sources that I admire—oriental architecture, Art Nouveau and the Arts and Crafts movement. And they were able to unify these separate elements and a myriad of highly wrought details to produce cohesive designs in which line and form, texture and color work to make one thing out of many.

The house that I bought (shown in the photo on p. 87) was designed and built in 1906 for Dr. William T. Bolton, a Pasadena physician, who died before the house was completed. It was bought and sold many times in the following years. Among the early owners were the Culbertson sisters, for whom the Greenes had designed another home a couple of miles away. The Culbertsons' shrinking inheritance forced them to move from their 10,000-sq. ft. home in Pasadena's Oak Knoll area to the smaller Bolton house (about 5,000 sq. ft. at the time) in the Orange Grove area.

The Culbertson sisters commissioned the Greene brothers to make a number of changes in the house, which ultimately added about 1,000 sq. ft. and which considerably altered the interior. Some changes were good; others were not. The turret stairwell was added in the early 1920s by the Culbertson sisters. It was designed by Garrett Van Pelt, an architect who had earlier worked for the Greenes. The stairwell is similar to one of two that were part of the facade of an earlier house designed for the Culbertsons' father by the Greenes, and built a few blocks away on South Grand St. This addition was built by the original contractor, Peter Hall. The Culbertsons changed the entry hall and writing room, removing some interior brickwork because they thought it was too rustic. Even with all of these changes, the house remained representative of the Greenes' style during the long time the Culbertson sisters lived there.

But the house did not fare so well with subsequent owners. In 1952, it was purchased by people who began a series of transformations that all but effaced the original appearance and character of the house. The new owners went to great lengths to turn a California bungalow into a French Provincial mansion. Their alterations included painting white the cedar and mahogany woodwork, painting over the doors and bookmatched mahogany paneling, removing and discarding all of the Tiffany glass lanterns and oriental-style box beams, and walling over two fireplaces that featured Grueby tiles.

When I bought the house, it had not been used as a residence for many years. After suffering the indignities of French Provincialization, the house had been bought by a local college and used as a warehouse and book depository. Bad things happened inside and out during its sentence as a storage depot. The roof went without maintenance and was allowed to leak. The plumbing wasn't repaired and leaked so profusely that water had rotted large holes through which one could gaze from floor to floor. Carelessly operated hand trucks and forklifts left hundreds of dents, scratches and gouges in floors, walls and trim. Most realtors would have classified it as a "fixer-upper." Indeed, the process that followed is better described as a resurrection than a restoration. Only a dedicated preservationist or a damn fool would tackle such a project. I qualified on both counts.

Restoring the Bolton house was a challenge and an opportunity. Our principal challenge lay in the fact that most of the original interior features had been entirely stripped away. This left us with the task of determining what the original rooms looked like, and having done that, reproducing their various architectural features. Our principal opportunity was to be found in the freedom to reinterpret creatively and fill in gaps with our own designs.

We were fortunate in finding that several people were interested in the rebirth of the Bolton house. The most recent owner, who had held it for only a few months, had tracked down several sheets of the original blueprints at the Avery Library at Columbia University in New York, as well as copies of all the building permits for the house and its many remodelings. The Greene and Greene library at Gamble House (4 Westmoreland Place, Pasadena, Calif. 91103) provided us with several 1908-vintage photos—both exterior and interior shots like the one of the original dining room below. Other people volunteered bits and pieces of information. All in all, we had quite a lot to go on.

However, having a lot of information is not the same thing as knowing what to do with it. None of us who worked on the project had been involved in a restoration project of this scale before, and restoration is infinitely more time-consuming and difficult than either renovation or new construction. You have to undo all of the old accretions before you can do anything new. You are faced with having to solve innumerable small mysteries before the big jobs can begin. You have to map the electrical, water, sewerage and gas systems before you can do anything with them. You have to reconcile the original blueprints with a structure that has been substantially altered, sometimes in baffling ways. You have to decide which of the original features are worth reproducing and which are not, while still remain-

Built in 1906, Greene and Greene's Bolton house suffered neglect and abuse before it was restored by the author, who was able to coordinate the efforts of local builders and craftsmen so that the work they did was equal in quality to the original woodwork, tilework and plastering. In the original entry (p. 87, bottom), the downspouts were made to look like structural elements, lending visual support to the wide overhangs. Missing here is the engaged half-turret to the right of the entry (p. 87, top), a later addition designed for new owners by architect Garret Van Pelt.

The original dining room, with its now long-lost custom furniture, was paneled with wide pieces of Honduras mahogany plywood. Unable to find good-quality mahogany plywood in these widths, Ross used standard 4-ft. wide panels and trimmed between with battens.

From *Fine Homebuilding* magazine (October 1983) 17:28-34

The kitchen (above), which had been Early-Americanized in the 1950s—knotty-pine cabinets and trim, scalloped copper work and an island range—had to be completely gutted. The window area where the sink stood was removed and the wall there filled in (right), and treated between the tops of the cabinets and the header strip with subway tiles.

ing faithful to your perception of the architects' vision. And, not least important, you have to stay solvent.

Working to high standards—I began the project in a fit of organizational energy. I made flow charts, job descriptions, checklists, spec sheets and budget estimates. I had a wonderful time, but none of these devices seemed to make the job go more smoothly.

There was so much to do that one of my first temptations was to tackle too many tasks at once. We had crews stripping paint, removing the old roof, clearing off the yard, and doing rough carpentry. We had plumbers plumbing, electricians wiring, and dozens of tourists getting in the way and asking questions. Keeping everyone organized, supervised, supplied and paid was tough—at times it was impossible. In the first few months, we went through more than 60 gallons of commercial paint stripper and two heat guns; we used up reams of sandpaper, innumerable other supplies, and lots of patience and goodwill.

In retrospect, I know we'd have been better off with a smaller and more tightly organized crew. Large crews can get work done faster, but they make mistakes faster, too, and sometimes the net result is not positive. Through natural attrition and the outright firing of some workers, our crew gradually shrank to a manageable size, and the overall quality of our work steadily improved.

Certain styles of architecture and interior detailing can be duplicated, more or less, using standard materials, conventional construction techniques and competent workmanship. Not so with Greene and Greene designs. The effect of the whole depends entirely on the near-flawless execution of seemingly insignificant details. Meticulous attention must be paid to wood joinery, to the finish and texture of the expansive wood and plaster surfaces, to the correctly done stained-glass panels, and to many other carefully wrought minutiae. And all these little things, taken together, create the Craftsman look and feel. There's just no such thing as an uncraftsmanly Craftsman house. So it became my responsibility to encourage the highest possible standards of craftwork, within the constraints of my budget and schedule. I think this is the most important role among the many that an owner-builder can perform.

Encouraging quality is different from demanding it. This kind of encouragement involves making a milieu in which craftsmanship will be the natural result of creative freedom, commitment to an ideal, adequate time and a spirit of cooperation. What this means in practical terms is finding bright, talented people who take pride in their work, and who have the right "chemistry" with each other. Further, it means insisting that such people work on a time-and-materials basis. This last point is an important one. Many trade and craftspeople are capable of much better work than they ordinarily do. But they are so accustomed to competitive bidding and the need to work quickly that they simply won't believe it when you tell them you want a job done right, even if doing it right takes longer and costs more.

Even accomplished craftspeople sometimes need to be convinced that it's okay to do the best work they can. The best way I know of to do this is to insist that they work on a time-and-materials basis, and then to work beside them as much as possible. Working with the crew does three things. It protects you by allowing you to oversee work directly—you don't have to wonder about how something was done, or if it was done. It gives you a chance to communicate your standards, and it establishes a bond of friendship and trust, which is a key ingredient of craftsmanship.

Doing things this way lets you experiment. Time invested experimenting pays dividends in the long run. Our skim-coat technique saved us a bundle. We did several patches right on the wall, varying the formula until we got it right and then proceeding from there. By experimenting we found we could mate old plaster to new drywall and blend them invisibly with a skim coat. This was especially helpful when replacing old ceilings. We also tried various wood finishes and techniques until we arrived at a compromise between ease of application and a good end result.

However, ordinary techniques that work well in new construction don't always work so well in restoration. The freedom to experiment helped us to solve problems in innovative ways, and saved money in the long run.

In the first several months of work on the house, we reroofed it, rebuilt the mechanical systems and dealt with structural problems. Some previous modifications had left walls with insufficient shear strength, and some modified openings were over-spanned.

The kitchen—The first room in the house to get intense attention was the kitchen. There was general agreement among the crew members that this room transcended bad taste and was better described as an example of "anti-taste," a deliberate, demonic effort to make a pretty thing ugly. It was incomprehensible to us that it had been featured in a 1952 issue of Los Angeles' *Home Magazine* with descriptions of its maple and knotty-pine woodwork, its quilted and scalloped copperwork, early American light fixtures and hardware, cork-tile floor and steel casement windows (photo above left). We promptly gutted the room.

We installed a maple floor, one similar to the original. New wood-frame windows were put in place, and we laid up tile on the walls from the floor to a continuous header trim (photo above right). These tiled walls are serviceable, and they're characteristic of the Greenes' kitchens.

In rebuilding the kitchen, our goal was to recreate the style and feel of the original, while at the same time creating a new and functional space with modern amenities. We worked toward this goal in several ways. First, we used a cabinet design patterned after the original but in a different (and expanded) configuration. Both the upper and lower cabinets were modifications of the original design, but close enough to capture their look (photo, page 90). This original look was maintained through the faithful retention of many details both large and small: the use of wooden countertops, the elimination of under-the-counter toe space and the use of Craftsman-style joinery (dovetails, finger joints and pegged tenons).

All of the appliances were carefully selected

Construction photos: Ken Ross

The woodwork in the restored kitchen is mostly redwood, and the cabinets are done in a style consonant with the Greene brothers' sense of scale and detailing. The fronts of the dishwasher and refrigerator are paneled, and the countertops are solid wood. Recessed lighting above the work areas concentrates light where it's needed, instead of flooding the whole kitchen with excessive illumination.

and installed to work visually with the overall design. The dishwasher, compacter and refrigerator were wood-paneled in a way that wouldn't attract too much attention to themselves. The refrigerator was flush mounted into a wall. A microwave oven was concealed behind a tambour door. We chose a commercial range for its appropriate scale and for its functional appearance.

The lighting was carefully thought out so as *not* to provide uniform illumination, with the idea that all work and eating areas would be brightly lit, while other non-work areas would be less so. This adds a degree of drama to the lighting, and also avoids the uniformly bright illumination of most modern kitchens. The cumulative effect of all this attention to detail is a kitchen that is modern and functional, yet entirely in character with the original style and intent of the Greenes' design.

In reworking the kitchen, we used a variety of modern techniques rather than slavishly adhering to old ways. We finished our countertops, for example, with a high-quality, solvent-based polyurethane, and we used a variety of modern sealers and preservatives. We did the ceiling and upper walls with gypboard, double sheeted and attached with screws. We laid up the wall tile on a plywood base rather than on a cement bed. For this base we used ⅝-in. exterior plywood, secured to the studs with panel cement and ring-shank nails. Instead of using thin-set, we set the tile with type A mastic, a high-quality mastic that's much easier to use and very strong.

For lights, we used a combination of leaded-glass lanterns in the style of the originals, and discreet modern fixtures. Our 2x12 (net dimensions) ceiling joists allowed us to install

The living room, which had been French Provincialized by the same people who remodeled the kitchen, is here being stripped of wallpaper and other accretions. By a stroke of luck, the locations of the original box beams, lanterns, header trim and other features were clearly outlined under the wall coverings. These worked almost like templates for constructing these elements anew.

A copper header with trailing-vine overlay conceals rebar holes above the fireplace lintel.

deeply recessed, museum-style down lights. Equipped with a ribbed black baffle and bezels painted to match the ceiling color, these fixtures faded into the background while at the same time putting light just where we wanted it. Although some preservationists would frown on these techniques, we felt that in almost every case we realized substantial benefits, we saved money and we kept from vitiating the spirit of the Greene brothers' original design.

The living room—After we finished rebuilding the kitchen, we concentrated our efforts on the living room. About the only things left over from the original room were the windows. All of the architectural features had been stripped away in the same remodeling wave that created the early American kitchen. This room, though, had been redone in a French Provincial style. Even the original sand-finish plaster walls had been covered with canvas, and then papered with a floral print. To our delight, when this canvas was pulled down, it revealed not only the original surface and colors underneath but also the white, unpainted silhouettes of all the original box beams, header trim and other ornamentation, as seen in the photo above left. This allowed us to verify dimensions and details that were unclear from our incomplete set of blueprints. But, our most exciting discovery came when we removed the reproduction French Provincial mantelpiece and chipped a small hole through the plaster behind it. There, covered over for nearly 30 years, was the original fireplace, with its handmade Grueby tiles still in place. Apparently these 1-in. thick tiles had been too hard to remove, so the 1950s remodeling contractor just covered them over with wire lath and plaster.

After photographing the room and carefully measuring the silhouettes, we pulled down the plaster ceiling because it was unsound and unsafe. We renailed the original wood lath with blue nails and covered it with wire lath, which we attached with heavy-gauge ⅝-in. staples and a pneumatic gun. Then we replastered the ceiling and finished it to get a sand-finish texture. The walls were extensively patched, and then the entire room was skim-coated (except for the new ceiling).

This skim-coating technique (also known as veneer-plastering) worked very well, and we used it extensively throughout the house. This saved us the considerable expense of completely replastering walls that were unsightly but basically sound. Instead of applying commercially prepared skimcoat, we used ordinary premixed latex joint compound and sand, with a little extra water added to help speed troweling. Conventional plaster doesn't work well for skim coating and should not be used. Our joint-compound formula hasn't shown a single crack in the entire house, and it's been in place for about three years now.

We finished the skim-coated walls with one coat of oil-base sealer and one coat of exterior oil-base primer before applying two coats of high-quality latex paint. We modified the oil-base paints with Penetrol and the latex with Floetrol, both made by Flood Co. (Hudson, Ohio 44236). These additives retard the drying time, minimizing lap and roller marks, and give the dried paint a very slight sheen. We painted the room before installing any of the woodwork, thereby avoiding the necessity of cutting in with a brush around innumerable edges and corners.

Restoring the living-room fireplace and hearth turned out to be a major undertaking in itself. Uncovering the original fireplace tiles was really good news. The bad news was that they were covered with mortar and plaster and pierced by two rebar studs, which had been inserted in the masonry to support the plaster. The hearth was missing all its tiles, which had been thrown out and replaced with marble, and the original copper firebox header was gone.

Then, in another fantastic stroke of luck, a local collector told me he had enough matching Grueby tiles from a demolished house to replace my hearth. Again, this was the good news. The bad news was that these sturdy tiles were still attached, in one monolithic piece, to 8 in. of well-cured concrete.

To make a long story short, we used a hammer and chisel, a water-cooled diamond saw and a lot of sweat to liberate the tiles intact. The tiles were cleaned by repeated scrubbings with a mild muriatic-acid solution and brought back to their original patina with a rubdown using diluted tung oil. We solved the problem of the two stud holes made by the rebar by designing a new copper header with an Art Nouveau trailing-vine overlay (also of copper), which extended up over the tile at the appropriate places and simply covered up the holes (photo above right). We borrowed the trailing-vine motif from the leaded-glass transom light over our front door.

In restoring the living room, we used Port Orford cedar for all the trim, doors, box beams and windows (photo, page 92). This wood was a favorite of the Greenes', who selected it for its durability, workability and its clear grain. The wood was kiln dried at the mill and air dried inside the house for several months. Port Orford cedar is durable and rot resistant because of its high resin content, but

The restored living room, with its Grueby tile fireplace, stained-glass lanterns and built-in inglenook bench, is almost a copy of the original Greene and Greene living room. The only significant change is the built-in audio cabinet to the right of the fireplace.

it can exude this resin for a long time and did so, right through our meticulously applied finish, in spite of all our precautions.

We used Honduras mahogany, also a Greene and Greene favorite, for cabinets, benches and lanterns. This wood was a pleasure to work, and yielded exceptionally beautiful results.

We began reproducing the woodwork for the living room by building the box beams. These were large heavy structures of various sizes, with radiused edges, arranged in an overlapping oriental design. We built our larger beams with interior ribs, much like a canoe, and attached all of them to the ceiling by nailing them to cleats affixed to the joists with drywall screws. The beams were nailed from the side with a pneumatic gun, firing 2-in. finishing nails. The nail gun was invaluable because it enabled us to do high-quality finish work in close quarters without nicking the plaster on the ceiling or denting the wood.

Before nailing the beams in place, we scribed them to the ceiling, a tedious, time-consuming operation that required us to take them up and down numerous times. The beams had to be brought in through a window, lifted up, held in place using pads and 1-in. by 2-in. braces (used like go-bars), then marked, taken down, planed, sanded and put back up again. Ugh.

One design feature that's truly a Greene and Greene trademark is a continuous wood header circling the room and joined at intervals with a pegged scarf joint, like the one in the center photo on the next page. The header joins all the windows and doors, and organizes them into one plane. It also visually lowers the ceiling to a comfortable height. In addition, the header serves as an attachment point for the light fixtures (usually lanterns) and as a picture molding. We installed continuous headers in the living room (and throughout the house), and mortised all the upright trim members—door and window cases—to receive them. Indeed, wherever two pieces of wood came together in this room they were either scarf joined, finger joined, mortised and tenoned or lap joined.

The most time-consuming and difficult woodworking jobs in the living room were the construction of nine wood-and-glass lanterns, an enclosed bookcase, and an inglenook bench. The Japanese-inspired lanterns averaged over 30 separate parts each, and required much meticulous care in construction (photo and drawing, top of facing page). These characteristic Greene and Greene style lanterns were attached to an inverted L-shaped bracket, which notched over the header trim and was screwed to both it and the wall below. The screws were recessed in square holes and then hidden behind ebony pegs. This attachment detail is used throughout the house. We don't know exactly how Peter Hall, the Greenes' craftsman builder, executed this detail. We did it by cutting the holes with a square, hollow-chisel mortiser and then tapping in a tapered oversized peg. The pegs were made from a long stick of ebony, milled to the proper size and then sliced like bread; then each one was tapered on a belt sander. Numerous small details like these pegs give a restoration job the stamp of authenticity.

The enclosed bookcase and inglenook bench are two of my favorite pieces. They were both designed along the general lines of the originals (we did not have details) and built out of Honduras mahogany. The framing members in both are joined with mortise and tenon. Panels are allowed to float free in their frames, so that wood movement across the grain of wide pieces wouldn't break the frame joints open. Some of the finger joints and tenons were locked with a screw, which in turn was set in a counterbore and hidden behind a peg. Since there is a gap under the peg, allowance is provided for the screw to move with the wood.

The dining room—By the time we finished the living room, we were really getting good at the kind of joinery and detailing that were such an integral part of Craftsman architecture. Work in the dining room, which was our next focus, went smoothly and quickly in spite of the many details. We had reached our stride. The original dining room had been paneled in Honduras mahogany veneer. In-

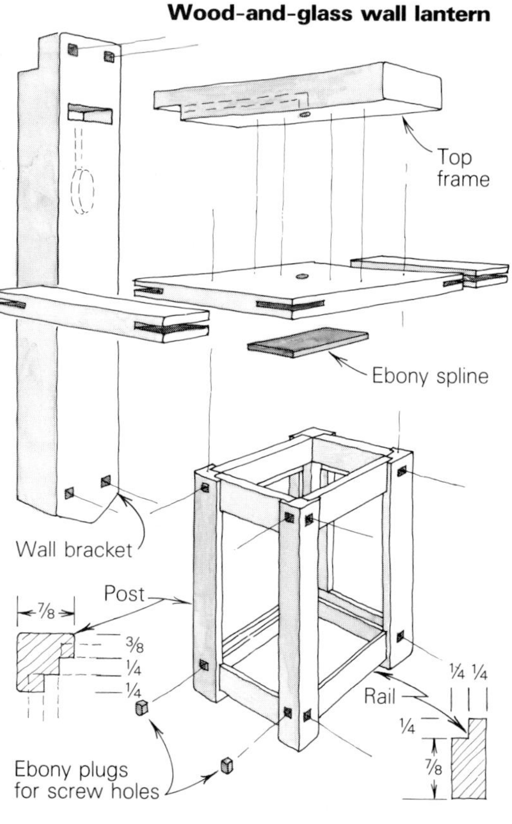

stead of trying to reproduce the veneer or buy high-quality mahogany plywood in these widths, we borrowed a pattern from the entry hall that used narrower sections of paneling separated by battens. For the panel sections we used ⅝-in. plywood with a Honduras mahogany face veneer. This was made to order for us by a local company and was surprisingly economical. This paneling was used with solid mahogany trim. We nailed the panels directly to the studs, being careful to align all nails so that they would be covered by the battens. The baseboard and continuous header were mortised to receive the battens, which were then finish-nailed to the panels. After filling, these nail holes were virtually invisible. In this fashion we achieved a tight batten-to-panel fit and also were assured that we would never have an open joint where the battens met the header and baseboard.

We also built a nook complete with window seats for this room. And we made nine more lanterns. The overhead lantern in this room is another one of my favorite pieces. It took almost two weeks to make. Even without the glass this lantern ended up weighing close to 50 lb. We attached it to the ceiling with four lag bolts screwed into the joists.

We achieved a beautiful color and patina on the paneling and other mahogany in this room and throughout the house by means of a technique we developed through experimentation. This technique involved first rubbing in a paste filler (if you don't use this, it will end up looking like lauan), followed by an application of clear sealer. Next came a pure green stain which we mixed ourselves. This stain, when applied over the reddish mahogany, produces a beautiful brown color. After the stain had dried, we applied three coats of a tung-oil base finish, rubbing between coats with 4-0 steel wool. The process is not as time-consuming as it sounds, and produces beautiful results. We used a tung-oil product throughout the house, except on areas that were subject to wear or friction of any kind. In these places we used polyurethane varnish and rubbed it out with steel wool or 3M pads until we got the desired patina. □

Made by Glen Stewart, a Pasadena woodworker, the Greene and Greene style wall lantern, top left, consists of over 30 parts, and is accented by ebony splines and square ebony plugs. The inverted L-shape mounting bracket is notched over the header strip and screwed to the wall, as shown in the drawing.

A signature of Greene and Greene detailing is the double-keyed scarf joint. It is sometimes used structurally, as in the beams, and sometimes used decoratively, as in the header trim, shown in the photo center left. To articulate the connection, the edges of the joining pieces are chamfered, and the keys are left standing proud. The header strip is mortised to house tenons on the vertical battens, and rabbeted to accept the paneling.

Far left, a pegged and chamfered finger joint connects pieces of header trim. Left, detail of the inglenook bench shows how the post is through-tenoned into the arm.

Photo: Paul B. Ohannesian

Before and after. **The author bought this bungalow because it needed a complete makeover and addition, but the tiny lot only had room in the rear (photo above). The 650-sq. ft. addition built in the rear (photo left) includes a studio, dining room, sun deck and stair tower. All existing clapboards were replaced to match the shingled addition.**

Bungalow Transformation

A stair-tower addition gives a small house new life

by Paul B. Ohannesian

In 1978, Vancouver was nearing the end of a period of stability in house prices. Sensing this, my wife Susan and I took the plunge to purchase our first house, a late 1920's bungalow on a 33-ft. by 111-ft. lot (photo above right). The house had a south-facing garden in the rear and was close to friends and within walking distance of Susan's work. It needed a lot of work, but this was an asset to me because I wanted the chance to remake the house totally. A secondary, but important, goal was to acquire the practical building experience that my university architectural education had not provided.

At just the right time, I met a young builder named Rae Gaylie. Gaylie did not then have a regular crew, and he was willing to take me on as his apprentice to rebuild the bungalow, inexperienced though I was. We agreed that he would be paid by the hour and that I would receive the benefit of any wholesale material prices he could obtain. I filed for a building permit and went shopping for a toolbelt and all-weather work clothes.

A clean sweep—We had purchased the house for its location and potential, but both building and property had many defects. Among the worst outside were an ugly carport consuming half the rear garden, a rudimentary sun deck in a hot, exposed location and inadequate drainage of storm water. Inside, there was no real entrance space, the floor plan was broken into small rooms, and routes through the house were circuitous. There was only one small, poorly finished bathroom, the kitchen did not have adequate counter and storage space, and there was no proper dining room—just a small eating nook off the kitchen. The stairs were steep and unsafe, wiring and plumbing were obsolete, the roof was worn out, and the walls were not insulated. To complete the picture, both exterior and interior finishes were of low quality and in poor condition.

We tried to correct as much as we could. I redesigned the existing floor plan to open it up, but the major transformation we made was a 650-sq. ft. addition containing an architectural studio on the lower floor, a dining room and a sun deck on the main floor and a stair tower (drawing, facing page).

Starting with the addition—If I had it to do again, the last month I would begin such a project would be November. I still have memories of starting work at 7:00 a.m. in the dark and in pouring rain. Our first job was to excavate for the footings, foundation wall and slab of the addition. Shortly after, I noticed water in the hole, even after several days of dry weather. We had dug into the hardpan below the subgrade runoff level. Lacking a connection to the storm sewer at the time, we dug a sump-pit and installed an electric pump, dumping

From *Fine Homebuilding* magazine (October 1989) 56:68-72

the water downgrade onto our front yard. The pump remained through four years of construction, a constant reminder of our vulnerability to a power outage, until we finally installed gravity drainage to the storm sewer. Fortunately, no blackout ever occurred.

The addition is 4 ft. below grade to provide a 10-ft. ceiling for the studio below the sun deck. We formed an 8-in. thick concrete foundation so that a 2x4 stud wall (made up mostly of windows) could bear on the inner four inches. The outer 3½ in. of the concrete wall were beveled 45° at the top, providing a strong visual base (drawings, next page). We hand-shaped the bevel of the circular stair-tower foundation wall with a homemade finishing jig as the concrete set (bottom drawing, next page).

To finish the inside of the walls, we built an insulated 2x4 stud wall spaced ¾ in. from the inside face of the concrete. The slab-on-grade floor received 2x2 sleepers at 16 in. o. c. The sleepers were individually leveled with plywood shims that were set on felt and nailed to the slab. We insulated between sleepers with 1½-in. rigid foam.

The remainder of the addition is stud-framed but with two extra twists. I like to see the "bones" of buildings, so I incorporated several exposed 4x12 fir beams. An 8-in. dia. Lodgepole pine column in the studio supports the junction of two such beams (left photo, p. 97). As in Japanese houses, it's a reminder of the living trees from which all wood comes.

To get adequate drainage on the sun deck, we shimmed the top of the floor joists to provide a slope to a drain. We applied a five-layer mopped built-up roof on ¾-in. plywood, then covered it with a protective layer of 90-lb. roll roofing. On top of this, we installed a pressure-treated cedar traffic-deck of spaced 2x4s that we screwed together in 2-ft. by 11-ft. units. These can be lifted for roof maintenance.

To the tower—The stair tower was one of those features that originally sounded like a good idea, was a nightmare to build, and ended up being a pretty good idea after all. A U-shaped stair would have been easier and cheaper to build than a circular stair, but the prominence of the tower, combined with my love of a challenge, dictated the final form chosen. We nailed and glued together 11 strips of 5⁄16-in. by 1½-in. plywood in a curve to make three plates. We bolted the first plate to the foundation, nailed up 16-ft. long studs at approximately 16 in. o. c. and nailed the second plate on the top of the studs. We sheathed the outside of the tower walls with two layers of 5⁄16-in. plywood, staggering the joints to stabilize the frame. We then hoisted two 4x12 fir beams for the tower roof framing into position with block and tackle and installed the top plate. Finally, we built the tower roof and installed the east-facing light scoop.

We tackled the stairs next. The 2x10 risers are supported on the outer wall and on an inner 2-ft. dia. core of studs. Treads of ¾-in. plywood are glued and screwed to the risers and supported by 2x2 ledgers. I fit fluorescent lamps vertically on two 2x4s at the very center of the core. I covered the inner core of the bearing studs with white Plexiglas diffusers and burlap-covered plywood panels, then covered the joints with fir strips (right photo, p. 97).

After insulating the tower, I had to decide how to finish the curved walls. Plaster on wire lath was my first choice, but I could find no one to do this work. I didn't trust drywall—I'd seen too many kinks and flat areas in such curved drywall work. As a compromise, we installed ¼-in. AC plywood—surface A exposed—with drywall screws, then taped and filled it as if it were drywall. We parged the surface with drywall compound, sanded it and applied a filler/primer and two coats of latex paint. The surface has held up very well over several years, although some horizontal joints have shifted slightly, causing minute cracks in the paint surface.

To make a handrail for the stair, I installed adjustable metal brackets with their top lugs ground off and fastened ¾-in. soft copper pipe to the brackets with sheet-metal screws, curving the pipe carefully. The railing is firm and has mellowed to the color of the stained woodwork.

Shingling—The existing painted wood-clapboard siding and poorly detailed trim were in sad condition, so there was no point in matching them on the addition. Besides, only a limited range of materials could be used on the tower's curved wall. I chose shingles partly for ease of installation by one person (myself), but also for their texture, reminiscent of California bungalows and local Queen Anne houses. I removed the clapboards from the existing walls before proceeding with shingling both old house and addition.

Common shingles were appropriate for most of the walls, but the end gables and the tower called for something more. I could not afford commercially available fancy butt shingles, so I hand-sorted 80 bundles of common shingles and pulled out all shingles between 6 in. and 7 in. wide. These were ripped on a table saw to a uniform 6-in. width, then cut in stacks of three to a 45° point.

This was to be a cozy house with a welcoming character, so I selected the exterior colors for their richness. All the common shingles are light brown. I precoated two thirds of the fancy butt shingles brown and one third red, using two coats of Benjamin Moore semi-transparent stain. Red also appears in minor accents on the south side in the tower window mullions and the sun deck

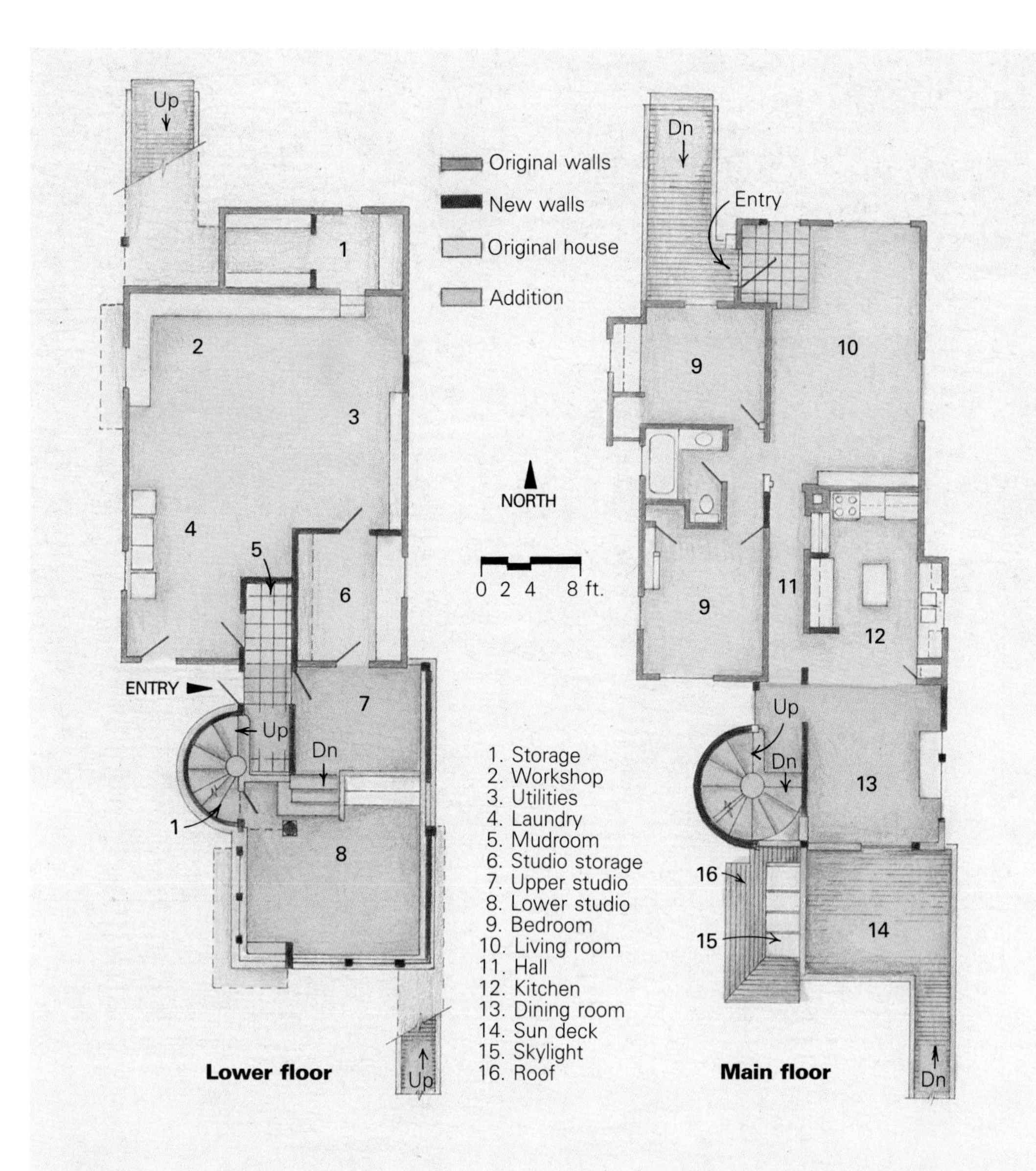

Drawings: Michael Mandarano

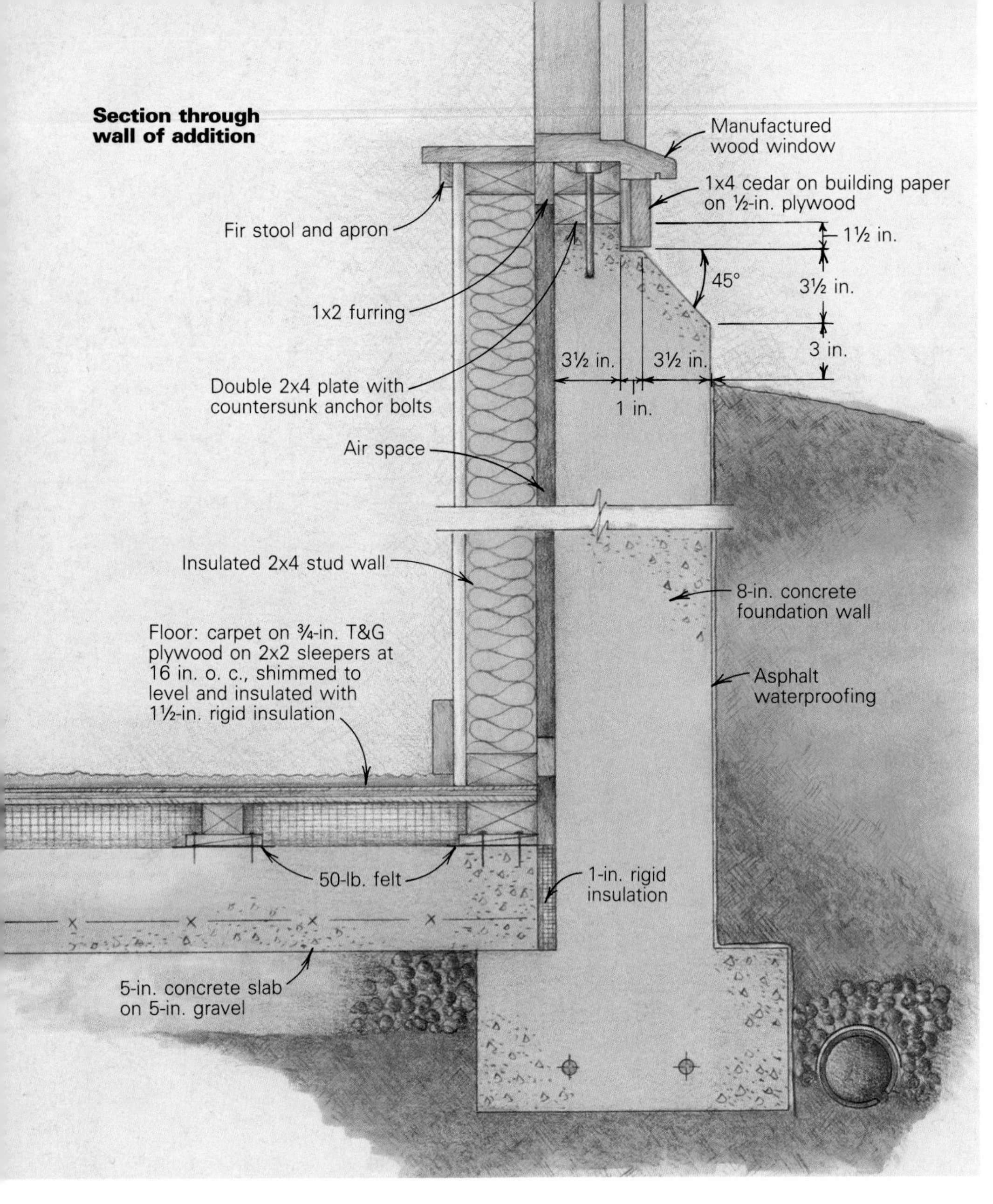

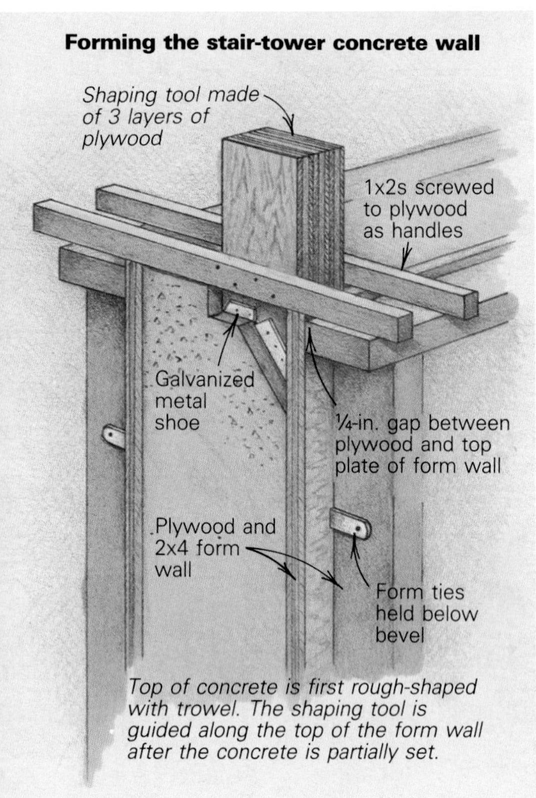

Top of concrete is first rough-shaped with trowel. The shaping tool is guided along the top of the form wall after the concrete is partially set.

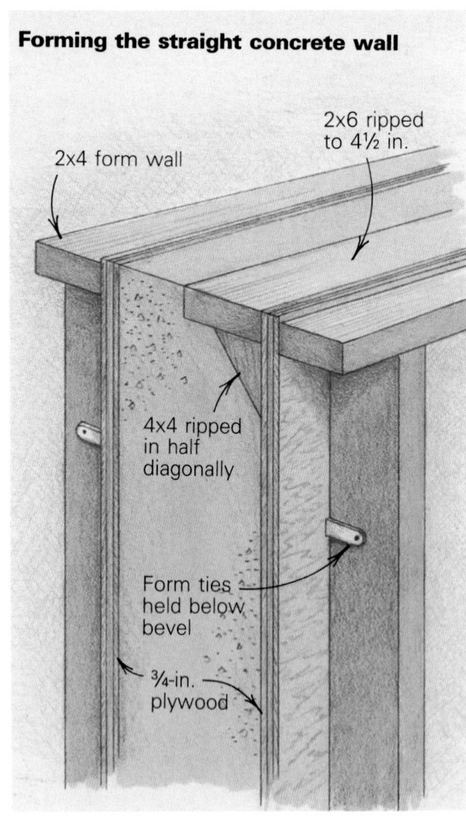

balustrade, and the new asphalt-shingle roof is a warm red-brown.

Chalklines could not be struck around the tower, so I leveled each course from the course before it. I laid the fancy shingles in a 2-brown-l-red-2-brown sequence to make a diamond pattern that starts as a narrow vertical stripe, widening as it approaches the ground and eventually girdling the full width of the semicircular tower wall (left photo, p. 94).

I had several options for controlling rainwater from the stair-tower roof. I originally chose a chain leader, which I used on the front of the house, but worried about its temptation to child climbers in the more isolated rear of the house. Susan and I had taken a trip to Europe before the project began and gargoyled cathedrals were still fresh in my memory. I carved a large block of cedar to a Viking-like profile, installed a plastic-pipe internal leader, and bolted the gargoyle to the wall at the top of the patterned stripe (top photo, p. 98). At first, water dribbled off its chin, so I installed a sheet-metal tongue to push the water several inches further. My English neighbor tells me that I am thus guilty of "rudery." No matter; it works.

Defeating the gloom—We replaced several of the existing windows. Unfortunately, our budget couldn't stretch to replace all of the single-glazed metal-frame units, so some remain to this day. I painted these to match the wood windows and trimmed them out identically. All new windows are wood sash, double-glazed. Operable units are either casements or awnings. The casements have proven quite effective in drawing breezes into the house.

In Vancouver, there are long periods of grey and rainy days when interior spaces become dark and gloomy. To counter the psychological effect of this, I incorporated a large skylight on the west side of the studio roof, adjacent to the sun deck. It's difficult to overstate the delight that this brings me on a daily basis; together with grade-level views out into the garden, the light from above removes any sense of working in a basement.

Windows, to my mind, should appear as more than mere holes in walls; they are the point of contact between occupants and the outdoors and deserve to be celebrated. For this reason, I followed the example of Greene and Greene and used exaggerated head casings on windows, and on doors as well. The 2x10 cedar headers overhang the 1x4 jamb casing (left photo, p. 94.). False sills of shaped 2x4 cedar, notched over the shingles below them, are likewise wide. Inside, windows and doors received 1x4 casing with the head casing carried out a few inches past the jamb trim; the corners were rounded. We stained the window sash inside and out to match the light brown shingles. All exterior casing is dark brown for contrast.

Floor to ceiling—One of the images I kept in mind when designing was the Japanese house, with its clean, uncluttered appear-

Part of the studio is below grade, but the windows and skylight prevent it from seeming like a basement room. In contrast to smooth surfaces such as the strip-cedar ceiling, an 8-in. lodgepole pine post carries one of several dark-stained rough 4x12 fir beams.

The circular staircase uses the curved outside wall and a center structure of individual studs as support. The author fit fluorescent strips between the studs and framed them with fir and Plexiglas. Curved walls are sheathed with ¼-in. plywood and parged with drywall compound. The stair rail is ¾-in. soft copper pipe fastened to metal brackets with sheet-metal screws.

ance and the controlled way that wood contrasts with white surfaces. To achieve this, we drywalled some ceilings and all walls, and painted them off-white. The walls of the stair tower are pale apricot to reflect daylight warmly. I ran 1x4 fir trim at door-head height throughout many rooms and used a 1x8 base that matches the existing base. The base and trim are stained a cedar color to give the walls the screen-like character of walls in Japanese houses.

I kept the floors simple and unified them with color—a light sandy beige in three different materials. Keeping the floors in a narrow color range makes a small house seem much larger. Most floors are carpeted wall-to-wall with a Berber-type carpet. Bathroom, kitchen and dining-room floors received cushioned, heat-fused sheet-vinyl flooring. We finished the upper and lower entry areas with 8x8 beige ceramic tile.

I've always been fascinated by what can be done in the ceiling plane to shape space and affect moods. Frank Lloyd Wright made use of this in his houses, but most architects miss the opportunity to make a space more than merely utilitarian by giving attention to its ceiling.

Some ceilings were simply painted drywall. In the kitchen, a previous owner had used oatmeal to texture the ceiling, which looked even worse than it sounds. Here, we resurfaced with ½-in. drywall. I painted the old plastered ceiling in the bathroom a dark color, then installed ¾-in. square cedar strips spaced ¾ in. apart. This not only provides an interesting surface, but it also absorbs condensation.

We surfaced some ceilings with cedar. Gaylie and I discovered a source of beautiful clear 1x3 smooth-faced T&G boards, many as long as 16 ft. I did a quick calculation of what we might need and purchased a large quantity. To resurface existing ceilings, I ripped a few boards into 1-in. furring strips, screwing these directly onto the old lath and plaster ceilings at 24 in. o. c. In the addition, we nailed the cedar directly to the framing. I applied no finish of any kind to the cedar; being overhead, it would not likely be soiled or damaged, and natural aging in the air has since turned it from a pink color to a rich tan. The cedar ceilings feel warm and protective and make a nice contrast to the white walls.

I took advantage of the fact that the dining room was all new and designed a sloped ceiling because gathering spaces, such as meal areas, require a space shaped to the "social bubble" of a group (bottom photo, next page). Originally, I intended to use the underside of the actual rafters on two sides as the finished nailing surface for the drywall, framing the remaining two sides to match the slope. In the end, however, we simply furred down the rafters, as the geometry of the roof framing was not quite even enough for finishing the ceiling.

I installed incandescent bulbs and stereo speakers above the cove, which is trimmed with unfinished clear 1x12 cedar. Initially the ceiling was painted apricot, but several years ago I asked my artist-brother John to airbrush a summer sky on the ceiling. In one corner, he painted a small winged cherub with the face of our newborn son Benjamin. As a finishing touch, we suspended our six-point candelabra from a toy hot-air balloon anchored to a hook in the center of the clouds.

Facing front—I dislike searching for street addresses at night, so I incorporated our house numbers into the front porch light. I applied 6-in. die-cut plastic numerals to a sheet of white Plexiglas and installed it on the face of a box that houses 100-watt lamps. The side of the box is a large wood return-air grille I had found in a scrap heap and backed with white Plexiglas. This face is attached with wood screws that can be re-

A leader drains through this gargoyle, carved from a block of cedar by the author. The sheet-metal tongue prevents rainwater from drooling down its chin.

The author made use of the new construction to give the dining room a coved and vaulted ceiling. His brother painted a sky several years after construction. The 8-in. base imitates the base found in the existing house, and the heavy trim on the windows follows an Arts and Crafts style.

moved for bulb replacement. The back, adjacent to the door of the house, is another white panel of plastic.

Hindsight—After a decade of living in and working on the house, I've made some observations about what I would or would not do again. I wish I had designed the stair tower so that each floor could be closed off from the tower. In summer, the tower functions efficiently as an air-moving stack: open a top window, crack one on the main floor and feel the breeze blow.

In the winter, however, much expensive heat is wasted through the stair tower. I've recently installed a curtain at the foot of the top run of stairs to prevent heat from escaping. This helps. A further measure would be to install a heat-recapture fan and duct, but this will have to wait until we upgrade our gravity-heat furnace to a forced-air unit. Another regret I have is that we did not put in a fireplace. Considering the other measures taken to assure a warm and cozy feeling in the house, the small extra cost of the fireplace would not have mattered much.

At the top of my list of hindsights is that I would not again enter into such an extensive project with an open-ended schedule and budget. Not getting firm quotes before beginning work resulted in several lengthy delays while more funds were earned or borrowed and a few periods of near-despair in a semicomplete house. As well, I would not keep my architecture office in a house under construction. I was, and still am, committed to combining my workplace and dwelling, but it was too much to expect that I could separate one from the other when everyday demands of directing trades distracted me from the equally demanding requirements of an architectural practice. Anyone planning a similar project should consider moving the office out of range of construction.

On the positive side, I am glad that I worked alongside a builder. Rae Gaylie taught me a great deal, not only about technique but also about thinking ahead. "Frame like a finisher" was his motto. I now think that experience in the field should be mandatory in architectural education. An architect can never foresee on a drawing board all of the fine-tuning that will occur during construction—especially on a renovation project—but having building experience will help close the gap between the mind that conceives and the hand that produces.

I have long believed that building a house is like the workings of a storage battery—you put energy into it by designing and building with care, and then for a long time afterward that energy comes back in small but steady doses of pleasure. That hypothesis was tested and proved true for me in the building of my own house. □

Paul B. Ohannesian practices architecture in Vancouver, British Columbia. Photos by John Fulker except where noted.

Top photo: Paul B. Ohannesian

Laguna Beach Remodel

Behind the restrained, street-level façade hides a rock-rimmed swimming pool and a house full of Craftsman details

by Patrick Sheridan

If imitation is the highest form of praise, then the Greene brothers are surely my architectural heroes. Simply put, I love the architecture of Charles and Henry Greene and continually use elements of their designs in my own work. For me a visit to the Gamble House (see sidebar, p. 103), arguable their finest work, is nearly a religious experience. But in addition to my liking the layered woodwork, finger joints and cloud-lift details, Craftsman bungalow architecture, and specifically the work of the Greene brothers, was designed for this latitude. The porches and courtyards that relate to interior living space and the broad overhanging eaves are well suited to southern California.

When I met Lee Sandler and Kathleen Farinacci, they lived in an unpretentious 1,450-sq. ft. house located in a gate-guarded beach community known as Three Arch Bay. Their one-story, 1930's stucco house sat on the side of a hill, below street level. Initially they wanted two things: to eliminate the steep driveway and to remodel the master bathroom so that it would be larger and more convenient. In the end we rebuilt the whole house, demolishing everything but two exterior walls and one third of the existing roof, adding a second floor and a total of 1,324 sq. ft. of living space.

Eliminating the steep driveway meant putting the garage at street level (photo above), which essentially meant adding a second floor to the house. This gave us an opportunity for an entry space on both levels that would tie the old and the new together. Although we remodeled the whole house, this two-story entry and the pool area at the base of it were the focal points of the project.

River rocks around the lagoon—The two main outdoor areas are the courtyard at the rear of the house, which already had a wonderful deck with a panoramic view of the Pacific, and an area at the front that sloped down and afforded little privacy. We decided to leave the rear as it was and concentrate on better use of the "front yard."

We decided that if the property was ever to have a pool, now would be the time to integrate it into the total package. Because of the prominent space the pool must occupy in the front, with half the house opening onto it, and because the main entrance to the house would lead people past it, the pool had to be something special (photo next page).

The Greenes used river rock in much of the hardscape (non-organic material, generally masonry, that is used to enhance landscape) surrounding their bungalows. We did the same, using river rock for perimeter walls, pilasters, retaining walls, planters and even seating. From a local quarry, we obtained over 50 tons of good-quality rock for just under $10 per ton, delivered.

For good rock work of this kind, the stone must be of a uniform composition and color. This can be accomplished by getting all the

From *Fine Homebuilding* magazine (Spring 1990) 59:83-87

A tropical paradise lies hidden behind the front gate. The swimming pool is over 10 ft. deep to allow diving from the river-rock wall that surrounds it. Over 50 tons of rock were used on the project. The mortar was dyed black and raked deeply between each stone. Scattered around the house, five sets of French doors (like the one shown above) help bring the outdoors in.

rock from the same quarry. Also, the faces and corners of the rock work must be as flat, true and as tight as possible. The finished product should look as though it were stacked without mortar. To this end, the masons mixed lampblack in the mortar and raked the joints between each stone and brick so the final effect was a shadow line only between the stones and brick. Although they approve of the finished product, my masons truly hate the lampblack because it stains like ink and gets on everything.

Some of the larger stones weighed a ton or more and were set with a crane; they were positioned to provide bench-like seating at poolside. The rock wall that surrounds three-quarters of the pool is actually a veneer over a Gunite retaining wall, which has a very deep shelf at the water line on which the larger stones were set.

At the highest point of the retaining wall is a pond that serves as a reservoir for the waterfall. The goal here was to create a sheet fall behind which a couple could sit. We built a seat behind the falls 6 in. below the water line of the main pool, and installed a light in the seat to illuminate the waterfall at night. Wanting the effect of a tropical lagoon, we trowelled the sides and bottom of the pool with black plaster, which also provides a significant amount of solar gain.

The rock retaining wall also satisfied one final design requirement. In college Lee had been a competitive diver, and he wanted a pool with as many diving opportunities as possible. We made the top of the retaining wall broad enough to stand on and dug the pool over 10 ft. deep.

Entry gate—Above the pool at street level, two large pilasters mark the entry while supporting the entry gate and pergola roof. The gate design was adapted from a gate at the William R. Thorsen house in Berkeley, designed by Greene & Greene in 1909.

We built the gate on site out of Honduras mahogany. The stiles are mortised into the rails, and the center panel is composed of 15 vertical pieces. The original design had seven vertical pieces with open spaces between them, but our need for privacy dictated that the spaces be filled, so we used an exposed mahogany spline, stained black, to act as a "shadow line" between the major pieces. All joints were glued with two-part marine epoxy.

Nearly as wide as it is tall, the gate is heavy, and there is no jamb where we could have fastened conventional hinges. We fabricated our own floor-mounted, center-pivot hinge. On the hinge side of the opening, we drilled a hole through one of the brick pavers and into the concrete sub-slab. Into the hole, we placed a ⅝-in. i. d., stainless-steel sleeve with the bottom welded closed. After packing the sleeve with some marine-grade grease, we dropped a stainless-steel ⅝-in. ball bearing into it, then again inserted more grease, then inserted a 12-in. long, ⅝-in. o. d. pin 6 in. into the bottom of the gate on the pivot side. The protruding end of the pin was dropped into the sleeve and serves as a pivot for the gate. For the top of the gate, we fabricated a steel strap and embedded it in the pilaster with about 4 in. sticking out. A pin installed in the top of the gate extends through a hole in the strap.

The ultimate staircase—In order to bring off the Craftsman style successfully, each element, down to the smallest detail, must have three things: a good design (proportion is all important); excellent materials; and perfect craftsmanship. I provide the design...with a great deal of help from Chuck and Hank Greene. And I supply the material. Every lumberyard and hardwood outlet in my area knows how picky I can be (the trick to getting good wood is to know when the new shipment is due and then meet it). My finish carpenter, David Hanson, provides the craftsmanship.

The stairs in this project were patterned after the staircase at the Gamble House. For years I have incorporated details of this amazing work, but trying to reproduce what many feel is the single most impressive element of the house was a treat. The stairs bring together all the best elements of Craftsman-style architecture (photo facing page).

David cut the stair-stepped handrail from a single 6x12 Honduras mahogany timber. Before cutting it, though, he plowed the top edge on the table saw to produce a 2⅜-in. deep by ⅝-in. wide groove. Once the handrail was cut to its stair-step shape, the groove became a series of slots at the front of each "step." These slots were filled with rosewood inserts. Using routers and chisels, David then mortised the balusters into the underside of the rail.

We paneled the wall beneath the stairs with horizontal pieces of 1x8 mahogany, which are finger-jointed to the risers and punctuated with rectangular rosewood pegs. Responding to Kathleen's desire for a hidden compartment, we built the lower portion of the wall so that it slides back to reveal a storage area under the stairs (bottom photo).

We chose quartersawn oak flooring for this project because it's much more stable than standard oak flooring and you get wonderful tiger-tail patterns in the wood. Also, it seemed to harmonize with the other woods. The floors in the Gamble House were laid on a bias to the walls so that the imperfect lines of the oriental rugs (which are never square) wouldn't be accentuated by the flooring. Because similar rugs were to be used here, we also ran the floors at a 45° angle to the walls.

Finger joints and cloud lifts—The finger-joint detail was used extensively throughout the house. Rather than mitering outside corners of boards that met at right angles, we divided the intersecting pieces into thirds across their width. One piece was cut to form two fingers, the other to form one, then all edges were eased with a block sander and the fingers were allowed to protrude slightly past each other. All the outside corners on the baseboards, on the fireplace mantel, even on the exterior trim are finger-jointed.

We also carried the "cloud-lift" motif through the entire house. Used extensively by the Greenes, this detail gives a sense of lightness to everything it graces. More interesting than a straight line, a cloud lift is simply a line that is offset, like a camshaft or a railroad siding. All the doors, interior and exterior, as well as the garage door, were made on site and carry this detail. For the windows and French doors, we used the cloud lift as a single muntin (bottom right photo, next page).

Another Craftsman detail that we incorporated was the Z-splice in the 6x12 ridge beam, visible under the ridge skylight in the upstairs hallway (top photo, next page). This splice substantially increases the surface area of the joint and reduces the tendency of the two beams to twist at their junction point. We used a 14-in. Makita beam saw with a carbide blade to cut the joint. The keepers are square stock set into the center of the splice and lock the adjoining pieces together.

The king post, which supports the ridge beam, sits directly under this Z-splice. In order to tie the king post to the ridge beam we could not merely mortise and tenon it as the Greenes did. My structural engineer insisted we use a ½-in. threaded rod bolted at both ends, bored through the center of the king post. A fake tenon, which appears to be an extension of the king post, actually caps the end of the threaded rod.

Although the staircase is patterned after the one in Greene & Greene's Gamble House, the hidden storage behind the sliding panel (right) is a new twist. The storage area was initially intended for wine, but the owners realized that vibration from the stair would be bad for wine.

Lights, lanterns and chains—All the exterior light fixtures, as well as some of the interior fixtures, were designed and fabricated in the Craftsman style. We had four copper lanterns fabricated, and for one of them, we borrowed another detail from the Gamble House. The lantern hanging in front of the garage has the house number in it.

In hanging these lanterns, with their delicate cut-copper panels and art-glass inserts, the chain needed to be welded rigid so these fragile fixtures would not beat themselves to death in the wind. Galvanized chain does not weld well, and I had a tough time finding non-galvanized chain. I eventually found it, though,

Moving the garage up to street level provided the opportunity for a two-story atrium. At the near end of the skylight you can see the Z-splice in the ridge beam. More than a decoration, this joint increases the surface area connecting the timbers and reduces the tendency of the beams to twist.

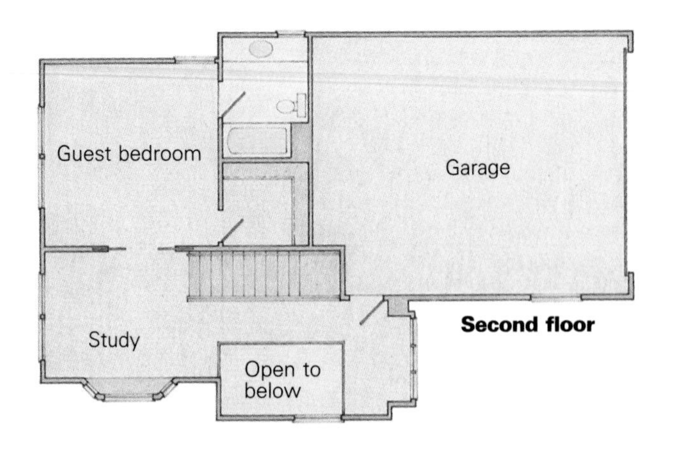

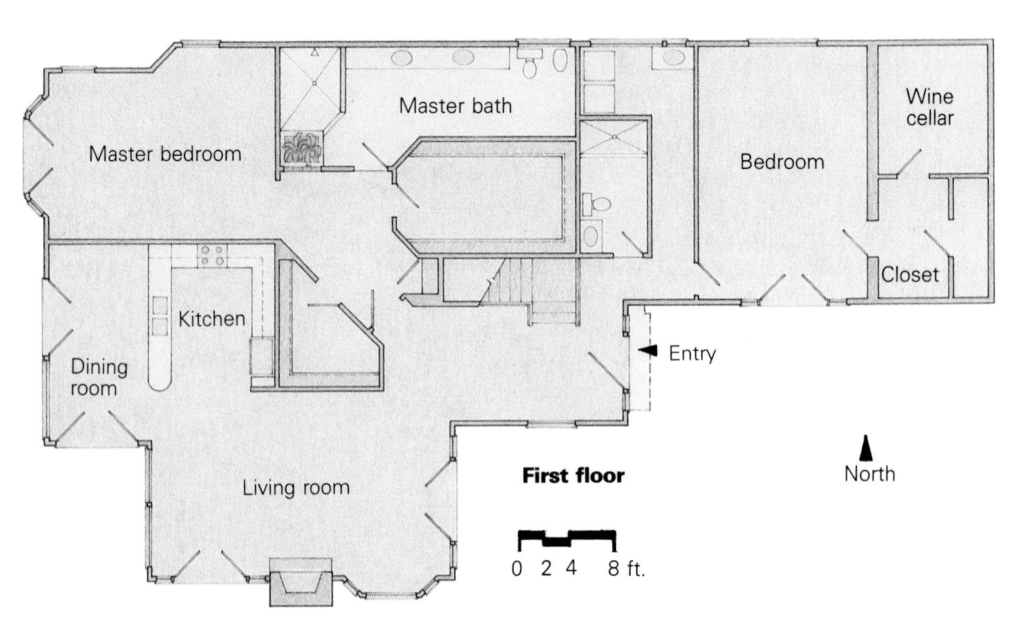

The hot towel bars are hardly a traditional Craftsman detail, but the Greene brothers might have appreciated the ingenuity of lacing the wall with stainless-steel pipes and routing the domestic hot water through them to heat towels.

Broad overhanging eaves and extensive outdoor spaces are among the elements that make Craftsman-style architecture well suited to the climate of southern California. Another Craftsman hallmark, the cloud lift, can be seen in the muntins of the windows and French doors.

and had four lengths welded rigid. Then I had them copper-plated, but left them unfinished to weather naturally.

Hot towel bars—It was gratifying to learn that we could blend hi-tech elements in with the Craftsman-style architecture. The house is equipped with three separate solar water-heating systems, each having its own panels: one for domestic hot water (with glazed panels), one for hot-tub heating (also with glazed panels), and a third system for the pool (with unglazed panels but with a much larger area).

Another hi-tech element we used are the hot towel bars. When Lee and Kathleen first expressed an interest in hot towel bars, I could find only electrically heated hot-oil bars, which seemed an unfortunate waste of energy. After talking to my plumber, I contacted a stanchion maker in a local boatyard who worked exclusively with stainless steel. From shop drawings I furnished, he fabricated two sets of bars: one four bars high and one six bars high, which, when connected to the hot-water loop of the domestic solar system, provide continuous hot towels without wasting energy.

The pipe is bent and welded stainless-steel ½-in. i. d. pipe, which snakes in and out of the wall in a continuous loop (bottom left photo, facing page), with a dielectric coupling at each end so the dissimilar metals, copper and stainless steel, do not eat away at each other. A hot-water circulating pump (Grundfos Pumps Corp., 2555 Clovis Ave., Clovis, Calif. 93612), which is placed at the solar storage tank, is needed to make this work, but is common in more expensive homes and also provides instant hot water to all outlets, irrespective of distance from the hot-water storage tank.

Greener and Greener—The project was started in late September of 1984, and we were filling the pool in time for the 4th of July 1985. Since that time other small refinements have been added, and I have had the pleasure of designing and building numerous pieces of Craftsman-style furniture. But what set this project apart from other bungalows I have done is that the clients had the intelligence and ability to commission the details that are all important to this type of architecture, or as David said near the end of the project, "It's getting Greener and Greener." □

Patrick Sheridan is a builder in Laguna Beach, California.

The Gamble House

When David and Mary Gamble, of the Proctor & Gamble soap empire, decided to establish permanent residence in Pasadena, they could have engaged any architect, as cost was surely not an issue. Fortunately for those of us who appreciate Craftsman-style architecture, the Gambles chose the relatively young firm of Greene and Greene.

Charles and Henry Greene attended the first manual training school in the U. S., where, in addition to their traditional high-school education, they studied woodworking, metalworking and toolmaking. This training had a profound influence on their architectural practice. From hands-on experience, they knew what good craftsmanship was and would accept nothing less.

Mary Gamble and Charles Greene worked out the basic plan, and on March 7, 1908, ground was broken. The Gambles were so confident of the Greenes' integrity and talent that they immediately left for a six-month trip to the Orient. The house was completed four months after the Gambles returned.

From the street, the dominant features are the bold horizontal lines so characteristic of the Greene's work (photo above). The broad overhanging eaves with their hand-carved rafter tails cast ever-changing patterns over the shingled siding. The porches and terraces, which were so carefully integrated into the rest of the structure, accentuate its horizontal lines.

The Greenes designed and fabricated not only the structure, but the furniture, light fixtures, carpets, place settings (including the silverware), picture frames and hardware. So complete was their attention to detail that a nook was built into the pantry to accept the extra leaves for the dining room table. As Henry Greene wrote, "The whole construction was carefully thought out, and there was a reason for every detail. The idea was to eliminate everything unnecessary to make the whole as direct and simple as possible, but always with the beautiful in mind as the final goal."

Seventeen different species of wood were used, including teak, mahogany, Port Orford cedar and oak, their varying grains and colors all carefully blended, and all glowing with a hand-rubbed finish. But it is the craftsmanship of the joinery itself that is the most amazing. Wherever two boards met, the connection was not hidden but celebrated and made part the design. The door-high horizontal trim, for example, that rings most interior rooms was scarf-jointed with Z-splices.

It is also gratifying to note that after 80 years the doors still hang correctly, the joints in the woodwork are still perfect, the furniture has never needed repair or restoration.

In their quest for excellence the Greenes eventually put together an enviable crew of craftsmen. But in the early years, the high standards demanded by the Greenes meant that subcontractors would often have to tear out work two and three times. Word quickly spread, and subs routinely doubled or tripled their bids. In 1905, however, the Greene brothers met the Hall brothers.

Peter and John Hall were born in Stockholm, Sweden, and neither had any formal training in woodworking. They were self-taught craftsmen. Their work was so fine, however, that within a few months of meeting the Greenes, Peter Hall opened a carpentry shop in Pasadena specially equipped to handle the designs coming from the Greenes' drawing boards. Most of the furnishings and much of all the millwork for the Gamble House came from this shop.

The Gamble House (4 Westmoreland Pl., Pasadena, Calif. 91103-3593) is now a museum, open to the public for tours Thursday through Sunday, noon to 3 p. m. —*P. S.*

Adapting the Japanese House

Eastern details, American function

by Len Brackett

After I completed my carpentry apprenticeship with temple builders in Japan, my wife Biva and I settled in the forested foothills of the Sierra Nevada Mountains in northern California. One of the reasons we chose to settle there is that close by are some of the last great stands of timber in the world—first-growth redwood, cedar, fir and pine. These precious trees are absolutely essential in the construction of *Sukiya* style buildings, where they are used to make the timber frames and the interior finish.

As I was building a new house for my family, I set up a nearby shop where my crew and I could first mill trees into dimensioned lumber and then cut the wood frames for Japanese-style houses commissioned by our far-flung clientele. Since Japanese houses are timber-framed, much of the work that goes into one can take place away from the building site. My crew and I spend months milling and cutting the parts for a frame that will eventually be assembled in a few days on some distant lot.

One day we got a call from a man on the East Coast who wanted us to build a westernized version of a Japanese house for a site in Woodstock, N. Y. Like most Americans, he wasn't particularly interested in living like a traditional Japanese, sitting on *tatami* mats on the floor with his shoes off. He wanted furniture, and a hardwood floor to put it on. So we designed a house of about 3,000 sq. ft. for him, which is very large by Japanese standards. The floor plan resembles an American split-level ranch house, with three bedrooms and two bathrooms over a garage and a general-purpose room at one end of the house (drawing, p. 106). At the opposite end, the living and dining area merges with the kitchen in an L-shaped plan.

Blueprints in hand, we met with the Woodstock building officials and got the necessary permits. Then we spent nine months shaping and finishing the parts—right down to the western red cedar ceiling panels. This is more offsite work than we usually do, but because Woodstock is a continent away from our shop, we wanted to complete as much of the house as possible before we loaded it into a container for its trip to the Catskills.

Just before we started to pack it up, however, our client called to say his plans had changed. A new marriage and a busy career had conspired to keep him in the city. So we pondered our options as we regarded our container-load of premium lumber, cut and ready to go.

Hideaways. **In traditional Japanese homes, the *engawa* is a buffer zone between indoors and outdoors, and its floor is slightly below that of the interior. In Brackett's westernized version (above), the floors of the *engawas* are raised to bench height and covered with *futons* to become cozy hideaways in the living area (facing page). *Shoji* line the walls of the living room and dining room, filtering the light and making the *engawas* private spaces. To the right of the curved beam, a laminated wood and glass light fixture runs the length of the room.**

We decided to build the house as a speculative venture on a suitable West Coast site. It was, after all, designed for an American, and we reasoned that there must be another one out there, living on the Pacific Rim, who could afford and appreciate this house.

Americanizing the Japanese house—Some American homes show the influence of Japanese architecture. Usually they have *shoji* (movable screens covered with rice paper or fiberglass) and roofs with exaggerated curves. But these are American houses that have had Japanese elements applied to them. We do the opposite. We start with a Japanese house and change it to reflect an American lifestyle. This way the elements that are crucial to a genuine Japanese house—the timber frame and its proportional elements, the natural materials and the traditional finishes—are integral parts of the structure.

As soon as we started to design a Japanese house that could accommodate furniture, we had a thorny issue on our hands. The proportions and relationships of the windows, doors, *engawas* (verandas) and ceilings were going to be off. In Japanese architecture, it is assumed that the line of sight will be close to the floor because one sits cross-legged on *tatami* mats. When you sit in a chair in a Japanese room, the space feels weird and off balance. The ceilings and windows are all too low.

Furthermore, traditional *engawas* are too small for furniture and below eye level for one sitting in a chair. If we didn't alter them in some coherent manner, they wouldn't fill the need that they address in a Japanese house: namely that of being an inside/outside space.

We began the design of this hybrid house by raising the floor of the *engawas* to bench height, and using the spaces underneath them for storage. Covered with a *futon* (Japanese bedding) and bolstered by a sloping pillow, our *engawas* became window seats that open completely to the outside (photo left). They can also be isolated by closing the exterior sliding-glass doors or the interior *shoji* (photo facing page). These spaces can be used for napping or reading, or even as a bed for overnight guests.

Once we had the *engawa* problem solved, the rest of the modifications followed suit. We

Photos pp. 104-107: Jeffery Westman

Hipped gable and shed roofs finished with copper shingles shelter this hybrid Japanese/American house on the Tiburon Peninsula in San Francisco Bay. Using native materials and the help of artisans trained in the ways of Japanese construction, builder Len Brackett and his crew spent 15 months attending to fastidious detailing throughout the house.

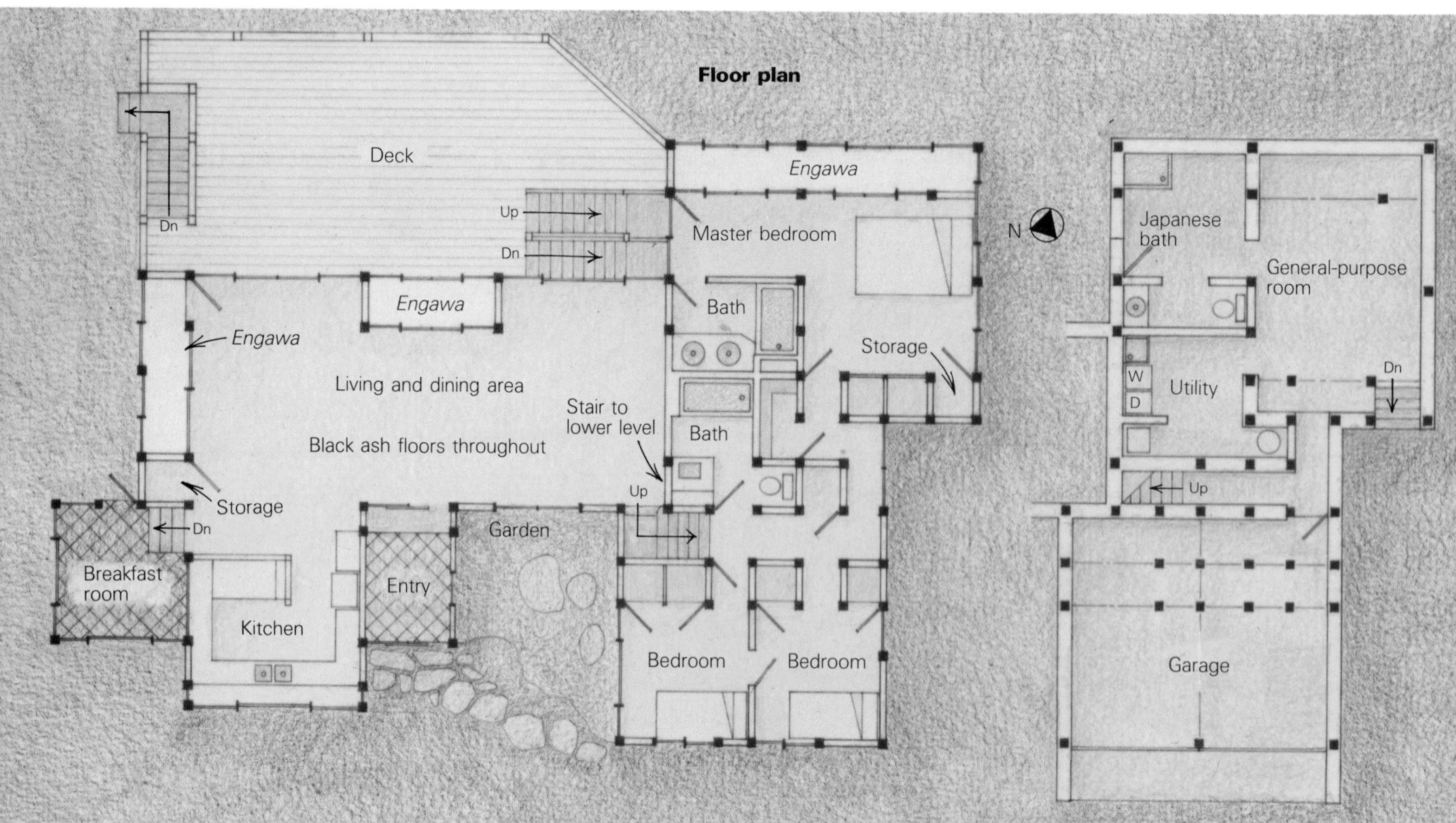

used stress-skin panels as infill between the posts for our exterior walls, which gave us the insulation we needed while preserving the look of the exposed posts. And by using rigid insulation on the roof, we were able to maintain the open ceilings that are part of the tradition of Japanese houses.

A Japanese kitchen is set up to cook miso and rice, and the heat source is a crescent-shaped woodburning hearth. Since this would appeal to only the most ardent Japanophile, we built a Western-style kitchen with a commercial range and a broad expanse of tile counter under a bay window (photo top right). Bay Area furniture maker Gene Agress built the kitchen cabinets out of maple and bubinga, and fitted them with Japanese hinges and pulls. Purists may shudder at our taking liberties with design traditions that go back 1,000 years, but my crew and I feel it is a challenge to make this beautiful architecture work for Westerners, and to build it largely with native American materials.

New state, new site—I spent about three months looking for a place to put the house. Every site I looked at had something amiss—access from the wrong side, slope going the wrong direction or a great view from the garage. I finally found the right lot in Tiburon, a peninsula that meanders into the northern lobe of San Francisco Bay.

But even this lot tested our determination. Tiburon has earth that's not well suited to support a structure, and before we could begin the foundation we had to remove 2,800 cu. yd. of soil to appease the engineers. Next came the engineer's prescription for the foundation. To create a stable footing for the house, we would have to build an enormous grade beam atop 28 concrete piers, each about 20 ft. deep.

We bit the bullet. After a couple of gulps and hard swallows, we hired a subcontractor to help us build a very expensive foundation. Our building inspector joked that all the compasses in the Bay Area were now aimed at the house, their magnetic pointers inexorably drawn to its mile-and-a-half of rebar.

Lumber—We adhere to the Japanese system of proportion and modules. In this scheme, which is called *kiwari,* all the framing members are directly related to the dimension of the posts. A squared post split into eight equal pieces becomes ceiling stringers. A post split into six equal pieces becomes ceiling molding. This system minimizes waste because if one has an extra piece, it can be reduced into usable constituents with no leftovers.

Since we end up needing oddball dimensions of lumber, we mill most of the wood that we use. For instance, the posts in this house are 4⅞ in. square and the top plate is a 5x10, which allows a slight overlap for a chamfer. Try finding a clear, dry sugar pine 5x10 at your local lumberyard sometime. Also, unless we mill the wood ourselves, we can't get the grain and figure we're looking for.

In all the houses we build, the posts are of Port Orford cedar *(hinooki,* in Japanese), the top plates and other spanning timbers are sugar

The kitchen cabinets (above) are made of maple and bubinga, and they continue the door details found throughout the rest of the house. For lighting, Brackett developed a process that bonds a thin veneer of wood to a glass or plastic lens, seen at right in a sconce in the bedroom hallway. Note the ceiling boards as they turn the corner.

pine from forests in the Sierra, and the ceilings are western red cedar or redwood.

The big spanning logs are mostly ponderosa pine and Douglas fir, and they are left exposed in several parts of the house. In the breakfast room, two of them with slightly different curves arch across the table—their weighty presence contrasting with the delicate, linear patterns of the walls. The curvy, elephant-trunk beam in the main room is an incense cedar.

All these logs came from the Tahoe National Forest, where they were doomed to be harvested along with everything else in a massive clear-cut. They were too curved to yield much lumber, so they would have been of little use as saw logs. We took them down in the spring, just as the sap was starting to run. At this time in their cycle they will give up their bark without a big struggle. We peel it off within a few hours after felling the trees, using a wooden bark spud to keep from marring the soft layer of wood under the bark.

All the milled wood surfaces that can be seen or felt in the house are hand planed to a glassy smoothness, and the logs are carefully washed with water and polished with soft cotton cloths. Except in ceremonial architecture (temples and shrines), the Japanese rarely finish architectural woodwork with sealer. Instead they use a hand plane to dress the wood (top right photo, p. 109), taking shavings as thin as 5⁄10000 in. Sanded wood picks up dirt if it's left unsealed, and its grain may eventually rise. But the grain in wood that

Drawing: Vince Babak

Timber connections. Mallets in hand, Brackett and his crew assemble the timber frame (left). To scarf together beams and purlins, they used complex connections called swallowtail joints (below left). Once the two pieces are driven together, the gap that appears at the center of the joint is filled with a pair of opposing wedges. Sugar pine top plates interlock above their post by way of dovetail joints (bottom). The diagonal channel is for the hip rafter. Seen from below, the finished assembly carries a broad eave of western red cedar boards and sugar pine and Port Orford cedar rafters (below).

Nicholas King

has been planed with a properly tuned and sharpened hand plane will not raise, even if it gets soaking wet. Planed wood also has a remarkable ability to resist staining. No oils, soaps, waxes or sealers are needed.

Even though planed wood will resist dirt, we take pains to protect the finished wood during construction. We paint the planed members with a thin wash of porcelain clay and water. The powdery residue soaks up fingerprints and other construction smudges, and comes off with a damp cloth at the end of the job.

Port Orford cedar, which is almost perfectly white when first planed, will become caramel colored in about 10 years. Within about 25 years it will be the color of milk chocolate, and in 50 years it will be almost black. All this time it gets shinier and smoother, if it receives the normal care of being wiped with a damp cloth from time to time. Many woods will act this way, not just Port Orford cedar.

Frame assembly, and a copper roof—Once we had the foundation work taken care of, the frame went together without incident. In three days all the bones were up, pounded into place with heavy wooden mallets (photo top left). Most of the intersections in the building are straightforward mortise-and-tenon or dovetail joints. Where sections of beams or purlins had to be scarfed together, we used swallowtail joints (photo above left) or goosenecks because they are excellent for resisting tension and bending stress. On swallowtails, the adjoining pieces have identical cuts on their ends. Once the two halves are aligned, opposing wedges are driven into the resulting gap in the center of the joint, driving the pieces together.

Where top plates meet atop corner posts, we use dovetail joints (photo above right). A diagonal channel cut into the top of the intersecting plates accepts the hip rafter. Also, the two plates are mortised on the underside to receive a tenon from the supporting post. The finished assembly (photo top right) is a crisp, uncluttered meeting of timbers that gives little hint of its complex innards.

We had originally planned for a western red cedar shingle roof, with about 2½ in. of the shingles exposed to weather, but local fire codes would not allow it. There are a lot of open spaces and woods in the immediate area, and the fire department didn't want burning shingles raining down all over the county in the event of fire. Tile? We didn't like what we found, and part of the point of this house was to construct it of native materials. Sheet metal? No. Slate? The roof wasn't steep enough. We ended up buying a ton of 8-oz. copper, and asked an old friend who is also a coppersmith to come from Japan to help us out.

He spent the first three weeks of his visit using a portable brake and a shear brought from Japan to cut and crimp the 3,000 interlocking copper shingles that went onto the roof. The shingles are about 6 in. wide and 2 ft. long, with flanges along all four sides. The top and right-hand flanges are folded up, and the bottom and left-hand flanges are folded down. The top flange of each shingle is secured to the roof decking with a pair of clips. When the next course of shingles is run, the flanges interlock with those of the preceding course (photo facing page, left), hiding the clips from the weather.

Working alone, it took our friend about two months to shingle the roof. As a crowning touch, he added copper gutters. Ornate splashguards catch the runoff at hips and valleys.

From *Fine Homebuilding* magazine (December 1987) 43:28-33

Copper roof. **Held in place by hidden clips, interlocking copper shingles cover the roof. At left, the roofer crimps mating shingle flanges.**

Natural finishes. **Brackett and his crew use Japanese hand planes to dress the milled surfaces of wooden elements (above). Wood finished in this manner will resist staining and raised grain without sealers or varnish. The interior walls (below) are finished with a layer of mud plaster made from dirt gathered at the site.**

Mud walls—In Japan, the walls of a traditional house are covered with plaster called *tsuchi kabe,* which is made from the local soil. The first layer includes straw for a binder, and it is spread over an armature of bamboo or wood lath affixed to the framing of the house. After it has dried, another mud mix is prepared with some sand in it. The sand helps to keep the plaster from cracking as it dries on the wall. Once the second coat is troweled on, the wall can be considered done.

It sounds easy in theory, but dirt is the most difficult plaster medium that I know of. Unlike most other plasters that can be reworked over and over again, a dirt plaster can't be fiddled with until it looks right. It has to be troweled on and left alone. Trowel marks are very difficult to hide unless the plasterer has a superb touch and an intuitive understanding of the material. If the mud is worked after the first application, shiny and dull spots emerge instead of the uniform, even texture of properly applied plaster. And since the ingredients sometimes come right out of the site, every batch behaves differently.

As we started on the grading for the foundation, I made a point of saving the topsoil for the small garden adjacent to the entryway. When our plasterer arrived from Japan and saw the pile of dirt for the garden, he was gratified. "Perfect color for the wall," he said. He really wanted that dirt, and I put up a fight for it, trying to interest him in the subsoil. No way! I could grow my garden in the subsoil, but the topsoil was going on the walls.

And so it did. Instead of a traditional lath backing, we used gypboard as a substrate for the mud finish. Our plasterer first applied a layer of stucco barely 1⁄16 in. thick to the gypboard for a ground. Then he troweled on the garden soil (photo above right). The finished plaster, roughly 1⁄4 in. thick, is one of the most vibrant surfaces I can imagine for a wall. It is chameleon-like, changing color throughout the day from almost green in the morning light to a deep rich brown at night.

Lighting—Other than recessed lights, we didn't have fixtures picked out when we began this house. Instead, we kept looking at catalogs and displays of fixtures at local showrooms, hoping to find something in keeping with the clean lines and subtle textures of the house. We never did, so we invented our own. By laminating thin veneers of wood to plastic or glass, we created durable, translucent shades for both incandescent and fluorescent lamps.

In passages such as the bedroom hallway, we installed sconces that have the veneer shades on them (bottom photo, p. 107). The wood veneer is thin enough to allow plenty of light to pass through it, yet thick enough to display the figure and color of the wood.

In the living room (photo p. 105), we installed long panels of the laminated shades to cover the fluorescent lamps above the *engawas.* When the lights are turned off, the fixtures appear to be solid wood. When they are on, they turn into a horizontal shaft of warm light that illuminates the complex ceiling.

Since we couldn't find anything else comparable to these fixtures, we patented the process and set up a company to manufacture the veneered panels. To find out more about them, contact Sloan-Miyasato (2 Henry Adams St., San Francisco, Calif. 94103). ☐

Len Brackett owns and operates East Wind Construction in Nevada City, Calif.

Cooperative Craftsmanship

Distinctive wood detailing, tile work, stained glass and carving in a California living room

by Ben Tarcher

Berkeley, California at the turn of the century was a carpenter's dream come true. An American hybrid design, the Craftsman style, had evolved, and hundreds of practitioners (affectionately dubbed "carpitects") took to the hills to fashion a community of peak-roofed houses out of seemingly limitless first-growth redwood and fir. Innovators like Bernard Maybeck (see pp. 74-81) and the Greene brothers in Southern California turned away from ornate Victorian designs, finding inspiration in such diverse styles as Japanese farmhouses and Swiss chalets. Houses became showcases for the warm and rustic tones of natural materials—stone, brick, tile and most of all, wood. Each project was a chance for artisans to display their talents. Woodworkers fashioned carefully detailed elements, from beams to staircases. Masons built massive fireplaces of stone, brick or concrete. Stained glass studios turned out leaded windows, and original lighting fixtures were created for the finest homes.

Today, the few empty homesites scattered about this rich architectural landscape offer a first-rate aesthetic challenge to designers and builders. My partner, Ed Hazzard, and I found such a site and on it built a modern home in the Craftsman tradition, employing talented artisans in a cooperative work atmosphere. Because

Exposed structural members, hallmarks of the Craftsman style, give the living room, left, a feeling of openness and warmth. The 4x12 cedar ridge beam is supported by three major trusses with king posts. Above is one of three crossed fish truss plates that were custom-made from 12-ga. steel. The plates tie the crossing chords of the trusses to the king post.

From *Fine Homebuilding* magazine (December 1981) 6:18-21

the home was a speculative venture on a difficult site, artistic considerations had to be tempered with practicality. The cheap land, labor and materials of the past that had allowed vast rooms and high ceilings no longer exist. To retain the sense of spaciousness found in earlier houses, we designed a soaring living room as a focal point. In the Craftsman spirit, simple, elegant detailing would distinguish this room.

Custom-built roof trusses—After we constructed 2x4 stud walls, a crane hoisted the 4x12 cedar ridge beam into place. We bolted pre-shaped cedar brackets to the beam where each truss would eventually go. We bolted 4x4 cedar king posts between the brackets and set 2x6 cedar rafters on either side of the posts. The rafters sandwich 1x6 cedar chords that pass through X-shaped, ¾-in. deep dadoes on opposite sides of the king posts. One chord on each side is continuous, the other is mitered and butted to its companion. The area between the chords contains spacers that stiffen the structure and serve as mounts for a light-track system.

We left the chords long, overhanging the wall plates. This way we could tension and straighten the long rafters by pulling on the chords. With the rafters straight, we temporarily nailed the chords to the rafters where they both cross the top wall plate. Finally we bolted them together and cut the chord ends flush with the rafter tops.

To connect the chords and the king posts, we used 12-ga. mild steel plates in the shape of crossing fish (detail photo, facing page) made by a local metal sculptor, Tatachook. To make them she first cut a template for the design pattern and made soapstone tracings on the sheet metal. She then cut the metal with a sheet-metal nibbler, leaving a margin to allow for grinding and finishing. She finally bored the holes for cap screws. We sandblasted the fish down to clean metal and then treated them with cold liquid gun-metal bluing to get a dark patina.

To achieve the strength of through-bolting and get a finished look without nuts showing on either side, we joined the king posts, chords, and brackets with ½-in. by 2-in. Allen-head cap screws. The king posts and spacers were pre-drilled to accept tapped steel sleeves (threaded from each end). The cap screws were then driven in from opposite sides until snug. The same method was used to hang the king posts from the cedar brackets on the ridge beam.

Interior finishes—The finished ceiling is 1x6 cedar shiplap siding nailed to the rafter tops. Wall surfaces are ½-in. drywall taped with taping compound troweled smooth over joints and nail holes. Pat Malley, a local plasterer, made the coved ceilings in the foyer. Ribs were cut from 2x12s to form the cove radius that joins the wall studs to the ceiling rafters. Where walls meet, we chamfered the corner ribs and covered the entire rib structure with expanded wire lath (drawing, right). Pat applied three coats of plaster over the lath on three consecutive days and tapered the plaster at the edges for the tape and compound to join plaster and drywall. We then sprayed walls and ceilings white.

To vary the natural colors in the living room

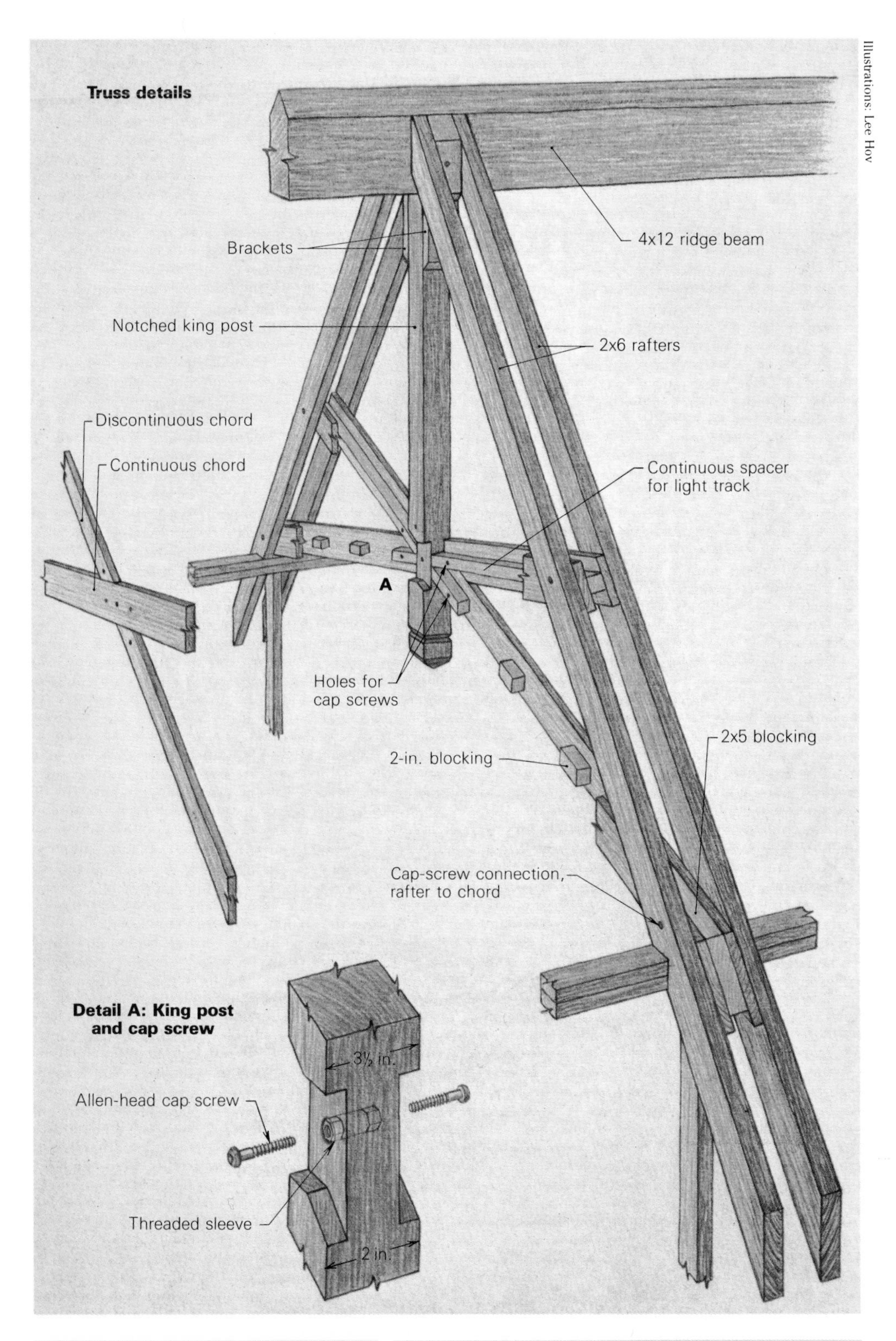

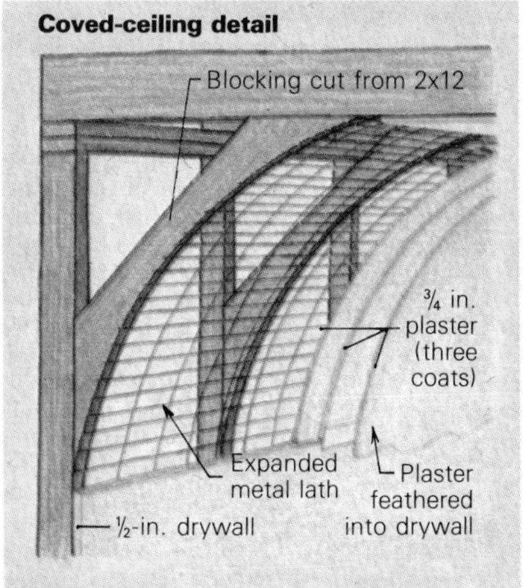

An inside corner of the coved ceiling. Curved ribs were cut from 2x12s, as at left. These ribs were covered with expanded metal lath, the base for three coats of plaster.

we worked a dark border into the oak strip floor, a design feature found in most Craftsman-style homes in this area. After painting the walls and installing the floors, we nailed up custom-milled, straight-grain redwood trim for windows, picture moldings, doors and baseboards. Before installing them, we sanded the pieces and oiled them with two coats of Varathane Plastic Oil, wiped dry after each coat.

Stair-rail construction—Woodworker Rick Magarian made the handrail at his shop from detailed drawings. He assembled it on the site. The finished stair is shown below; for construction details, see the drawing on the facing page. The channel-shaped handrail is a composite of three pieces—two sides and a top rail, which was shaped from 2x4 red oak. Its top edges were rounded over with a router, and two grooves were plowed along the underside of the rail to house the two sides. Rick coved the outer faces of the side pieces on the table saw, clamping a pair of auxiliary fences atop the table to feed the work over the blade at about 60°.

Rick bandsawed the curved sections for the landing handrail from solid blocks of red oak, then sanded them to shape with a 3-in. spindle sander. He steam-bent the side pieces to conform to the rail, and slotted the curved parts of the handrail by attaching a router to a pair of trammels. He joined the handrail section to the lower shaped section with two standard ⅜-in. by 3½-in. dowels at each union.

The square-headed balusters were ripped from 2x cedar stock and shaped on a jointer with stops on the infeed and outfeed tables. The center areas were relieved by carefully lowering the material over the rotating cutterhead and running it through to the outfeed stop. Each side of all three different-length balusters went through the jointer twice, with a ⅛-in. wood shim under the balusters for the first pass to avoid possible kickback, and to remove only ⅛ in. on the final pass, a light enough cut to eliminate tear-out and chipping. Sandpaper is glued on top of the shims so the shims won't slide; for an extra measure of safety, the shims could be taped to the balusters.

Rick jointer-shaped the newel posts from 6x6 cedar timbers, then notched and faceted them with a radial arm saw. With the newels nailed and bolted in place, he plugged the remaining holes. He then glued and screwed a maple ledger block to each newel. These blocks fit into a hollow in the end of the handrail and tie the rail and newel together once the balusters are set in place. When the newel was ready to receive the rail, Rick secured the balusters to the stair treads with dowels, using slow-drying glue. There are two balusters, one longer than the other, on each tread. In all there are three lengths of balusters, two for each tread and one for the horizontal landing sections of the rail.

In preparation for joining the rail to the balusters, Rick cut spacer blocks, which he nailed and glued to the underside of the railing at intervals to match the spacing of the balusters already in place on the treads and landing. An extra ⅛ in. was left between the blocks for maneuvering the balusters into place. He then applied glue to the contact points on the underside of the rail, slipped the rail over the maple block on the newel, and one by one worried the balusters into place under the rail. Then he glued the other coved handrail side into place and clamped the sides together. Finally, he glued and nailed oak cover blocks between each pair of balusters, and the balustrade was complete except for finishing—a sanding with 400-grit paper and two coats of Watco oil. It took Rick and our head carpenter, John Palms, 230 hours to build and install the stair.

To complete the Craftsman theme we hired tilemaker Jeff Bickner to make the tiles surrounding the fireplace and cabinetmaker Miles Karpilow to design, build and carve the fireplace mantle. My partner and I had already decided on the dolphin theme when we first talked with Miles. The dolphins, an ancient symbol of intelligence, freedom and the collective spirit of man, seemed perfect for our house.

And the collective spirit emerged intact at the project's conclusion. These craftsmen brought their experience and enthusiasm into this home and proved that the exemplary workmanship of the past can still thrive in the present. □

Ben Tarcher is a graduate architect and contractor in Berkeley, Calif.

It took two men 230 hours to build and install the stair. The handrail is red oak; its side pieces were coved out on a table saw, and the balusters were shaped from 2x cedar stock.

Windows, as at right, are bordered with leaded glass with a horizontal, undulating strip of colored glass at the top. Each piece of trim was custom-milled from straight-grain redwood and finished with Varathane Plastic Oil. The stained glass is the work of Bruce McLean. The fireplace, far right, features tiles by Jeff Bickner and a dolphin mantle by Miles Karpilow.

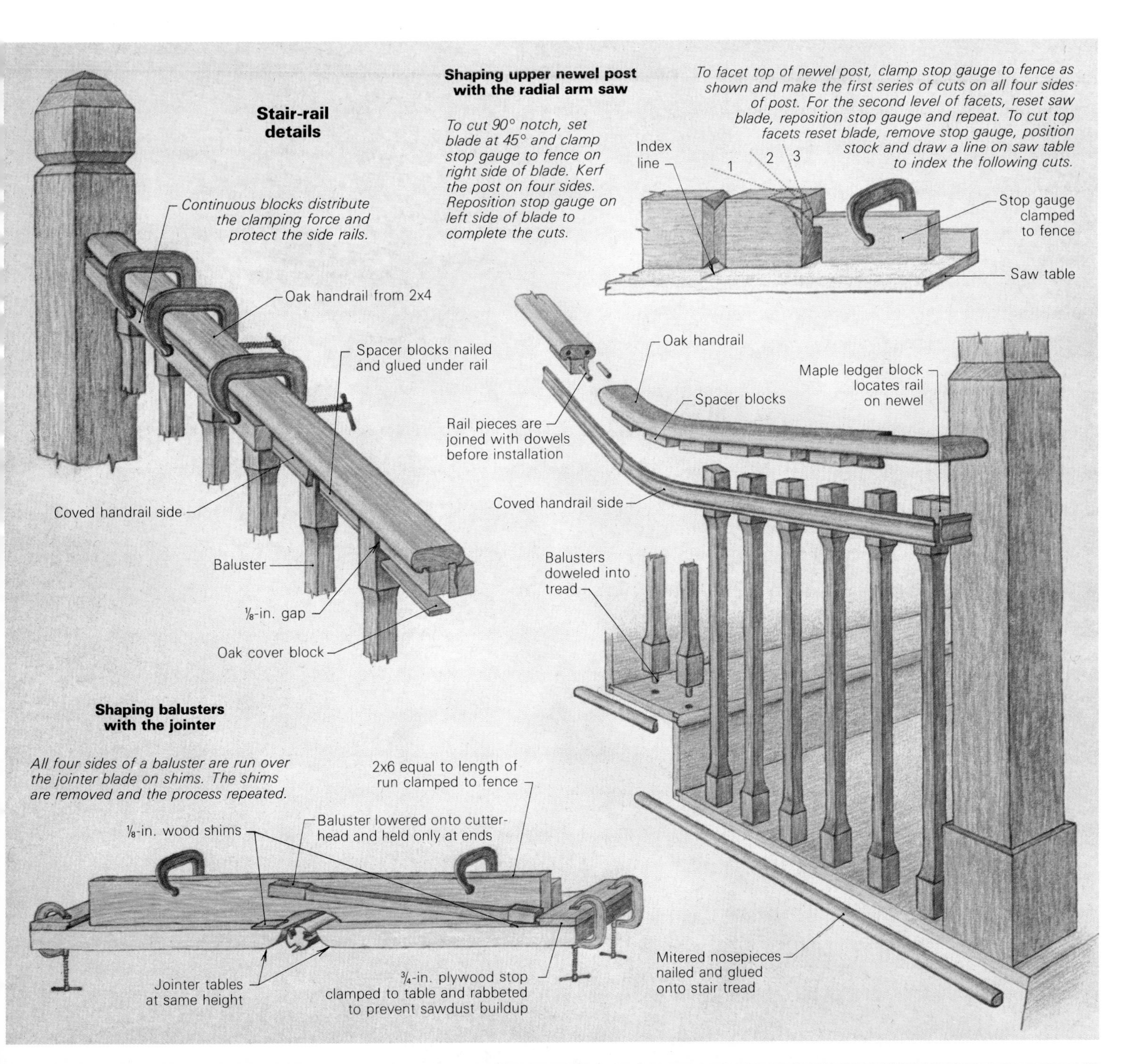

Stair-rail details
Continuous blocks distribute the clamping force and protect the side rails.
Oak handrail from 2x4
Spacer blocks nailed and glued under rail
Coved handrail side
Baluster
⅛-in. gap
Oak cover block
Shaping upper newel post with the radial arm saw
To cut 90° notch, set blade at 45° and clamp stop gauge to fence on right side of blade. Kerf the post on four sides. Reposition stop gauge on left side of blade to complete the cuts.
To facet top of newel post, clamp stop gauge to fence as shown and make the first series of cuts on all four sides of post. For the second level of facets, reset saw blade, reposition stop gauge and repeat. To cut top facets reset blade, remove stop gauge, position stock and draw a line on saw table to index the following cuts.
Index line
1
2
3
Stop gauge clamped to fence
Saw table
Oak handrail
Maple ledger block locates rail on newel
Spacer blocks
Rail pieces are joined with dowels before installation
Coved handrail side
Balusters doweled into tread
Shaping balusters with the jointer
All four sides of a baluster are run over the jointer blade on shims. The shims are removed and the process repeated.
2x6 equal to length of run clamped to fence
Baluster lowered onto cutter-head and held only at ends
⅛-in. wood shims
Jointer tables at same height
¾-in. plywood stop clamped to table and rabbeted to prevent sawdust buildup
Mitered nosepieces nailed and glued onto stair tread

0 4 8 12
Scale in feet
Pond
Bedroom
Entry
Laundry
Up
Guest suite
Dn
Hall
Kitchen
Living
Study
Dining
Lanai
Bath
Master bedroom
First-floor plan
Pool
Bedroom
Dn
Bath
Second-floor plan
N
Pacific Ocean

Hawaiian Retreat

An architect builds a subtropical getaway on an island paradise

by Charles Miller

Ralph Anderson is an architect who admits he likes plants better than buildings. When he recalls the homes that he has designed, many of them strike him as elaborate foils for landscaping. Lush courtyards, decks at tree level and views framed by carefully placed windows are to be expected. So it comes as no surprise that Anderson's own house, on the island of Kauai, is centered in a seaside garden with plenty of glass to let in the views.

Anderson and his wife Shirley live in Seattle, and they wanted a vacation house where they could escape the Northwest rainclouds. Kauai's subtropical sunshine and mild climate—it's almost always between 75°F and 85°F—offered the Andersons the contrast they were looking for. In scouting the island for a home site, they focused on the Poipu Beach area. Located on the leeward, southern tip of the island, Poipu enjoys constant breezes while ducking the brunt of the trade winds. Here morning and evening showers are a near certainty, but they're brief and often followed by rainbows.

At Poipu, the Andersons discovered an intriguing property that was up for sale. It was a 1.2-acre site that included a house built on a rocky point surrounded on three sides by ocean (photo facing page). The house was designed by Hawaii's most famous architect—the Russian emigré Vladimir Ossipoff. Vaguely reminiscent of a California ranch house, Ossipoff's building was long and narrow, with its southern facade broadside to the ocean. By the entry on its north side, there was a fresh-water fishpond with a waterfall fed by a recirculating pump.

Taken with the site, the Andersons bought the property. Ossipoff's house was good enough to save, although Anderson felt it didn't do the point justice. He tended the gardens and enclosed the entry porch, but secretly he wished he could tear the house down and start over. In November 1982, powerful winds generated by hurricane Iwa blew the roof off the house and flattened its walls. His wish granted, the architect got busy designing.

Facing page: The Anderson house sits on its stony point in a grove of coco palms. The plan and elevation drawings show the house's three distinct structures. The long hallway that connects these structures, above, is lined with columns of lava rock and topped with a low gabled roof.

Materials, siting and a plan—The point is solid rock—lava that cooled into a wrinkled, dark brown skin when it hit the sea eons ago. Lava rocks the size of volleyballs are strewn up and down the coast, and they are often used in local buildings for walls and columns. This concept appealed to Anderson, so he began thinking of the house in terms of a post-and-beam building. The posts would be steel-reinforced columns of mortar and lava rock. The beams would be web trusses, scissor trusses and solid timbers, all made of Douglas fir and anchored to the columns with heavy steel plates to resist the next hurricane. Privacy walls between the street and a nearby house would be solid rock or cedar siding, and the rest of the walls would be glass.

Zoning regulations have changed since the Ossipoff house was built in 1940. The new house could either occupy the same spot as the old one or move toward the road, farther from the ocean. Anderson chose the first option. For better views and some insurance against tsunamis (earthquake-generated waves) and wind-driven tides, Anderson raised the grade a few feet above the previous elevation.

Anderson's plan links three distinct structures (drawings below and on the facing page). The southern elevation shows the massing of the three buildings, which together evoke a neighborhood street scene with a church-like pavilion in the center.

The building to the west houses a suite for guests, and contains the kitchen and dining room. At the eastern extreme, a two-story wing

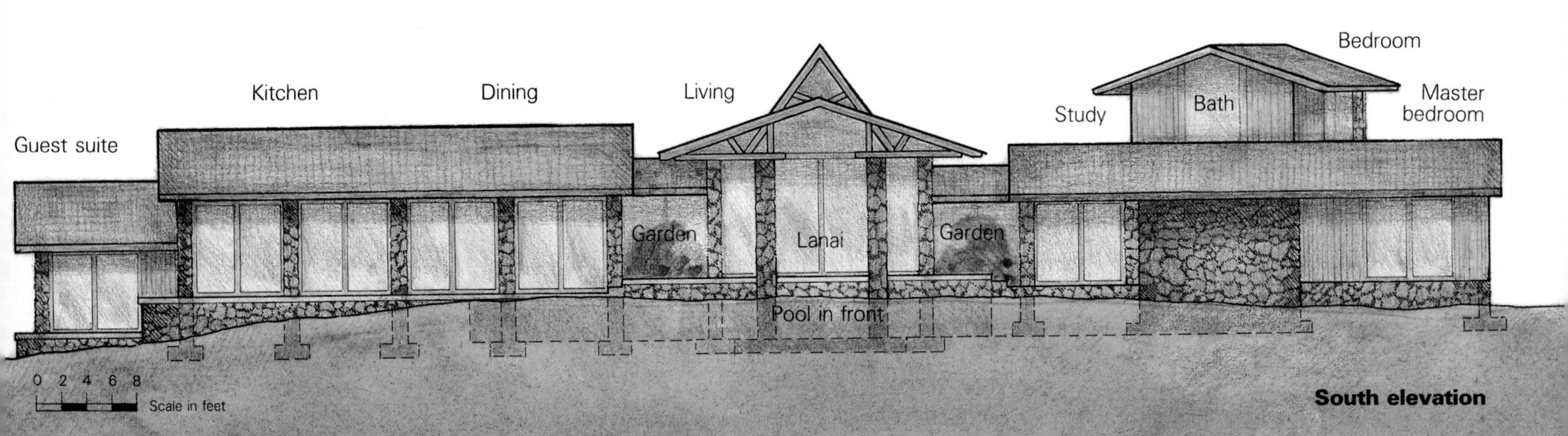

On the south side of the house, above, a cantilevered roof supported by lava-rock columns spreads its eaves over the lanai. Anderson's scissor-truss design (drawing, below) was inspired by the dual-pitched roofs found on many local buildings. Below, the roof emerges on the north side of the house, where it shelters the front door. Louvered panels fill the door and flank the entry, allowing cross ventilation through the living room. Insect screens on the inside keep out the bugs.

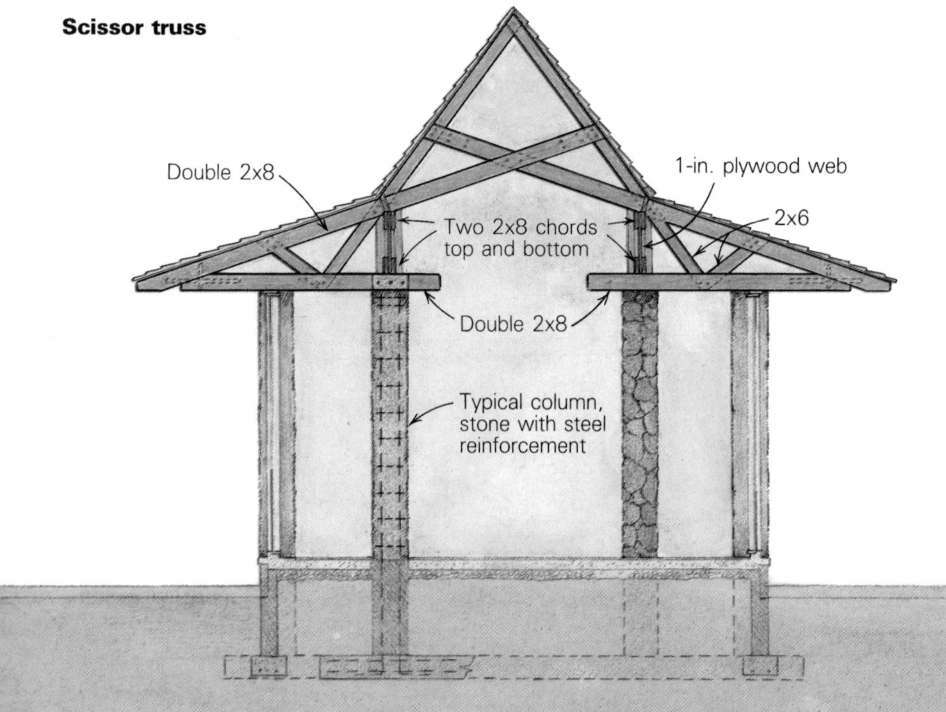

includes three bedrooms, with a bathroom for each, and a study/TV room. Connecting the far-flung ends of the house is a corridor nearly 90 ft. long (photo previous page). This hall, flanked on both sides by lava columns and topped with a low, gabled ceiling, imparts a cloistered, medieval serenity to a house that in every other respect sends the eye off to scan the horizon.

There are over forty sliding doors in the house. Virtually every exterior opening between the columns, which are mostly on 10-ft. centers, has a fixed window, a glass slider and a screen slider. Heating is not an issue (the house doesn't even have a heater or a fireplace), but cross ventilation is all-important. For most of the day, Anderson keeps the doors wide open.

Perpendicular to the long hall is another axis that begins at the entry and bisects the pavilion in the center of the house. A step through the front door reveals immediately the focus of this home—an indoor/outdoor living room topped with a complex and finely wrought roof structure, surrounded by plants and water (photo facing page). This room is isolated from the adjacent rooms by miniature gardens, 8 ft. wide, that are planted with ferns and fishtail palms. With the pond's waterfall and the distant breakers, the sound of falling water is ever present. In the view to the south, framed by lava columns, tubular turquoise waves break over a reef known by local surfers as the Acid Drop.

Roof form—Steep hipped roofs are common in Kauai, especially the kind known as bonnet roofs. Like a hat, these roofs have eaves of shallow slope (the brim) merging with a steeply angled gable (the crown) over the center of the house. In the subtropics, this type of roof has advantages: warm air exits through vents in the high ridge, setting up a convection loop that helps to cool the house; and the wide eaves keep the rain and hot sun off the walls.

Anderson put the double-pitched roof to work in the center of the house, using it to shelter the living room and lanai (photo top left). But rather than make it a hipped roof, Anderson used scissor trusses to make a dual-pitched gable roof (drawing, below left). The lower gable rises at a 4-in-12 slope, then meets a second roof plane that rises at a 16-in-12 slope to the ridge. On the south side, the scissor-truss eaves cantilever over the lanai like wings, protecting sunbathers from cloudbursts or intense midday rays.

The dual-pitched roof extends across the house like a mammoth extrusion to end above the north-side entry (photo above left). Anderson stopped the steeply pitched portion of the roof in line with the front door, and extended the lower roof to complete a single gable that shelters the steps. This keeps the roof from looking top-heavy, and it makes a good place for a clerestory to help illuminate the ceiling.

Ceiling detail—Anderson wanted the wood in his house to be reminiscent of the Arts and Crafts period. He wanted all exposed edges to

Facing page: From the front door, columns frame a view of the ocean.

From *Fine Homebuilding* magazine (Spring 1987) 38:78-83

The living room is flanked by small, palm-filled gardens, which isolate the room from the rest of the house and amplify its tropical setting. In this photo, one of the two web trusses that carry the complex series of scissor trusses can be seen atop the lava columns.

be rounded over, and he wanted the wood paneling to be sanded and oiled to heighten its grain. The architect was willing to pay for this level of finish—if he could find a crew to pull it off. John Carothers, a transplanted carpenter who learned his trade in California, turned out to be the right builder.

Carothers looks at the truss matrix that he and his crew built (photo above), and describes its construction as "grueling." "Every stick of wood in these trusses had to be handled eight or nine times, beginning with hand-picking each 2x6 and 2x8 at the lumberyard," he recalls. Because the scissor trusses interlock with so many neighboring components, they couldn't simply be assembled on the ground and then hoisted into place. Instead, they were glued and bolted together in place, one by one.

Carothers made precise patterns for each truss member. Once the parts were cut, the edges of the 2x stock were eased with ¼-in. roundover bits. Then the pieces went to the guest house, which had been set up as the sanding shop. There, three workers in dust masks, goggles and ear plugs sanded Douglas fir for months, burning out three belt sanders in the process.

No fasteners show in the scissor trusses. The ½-in. machine bolts at each intersection were counterbored, then capped with fir plugs. Tongue-and-groove 1x3 fir floorboards were used to panel the inside of the ceiling, followed by a layer of 6-mil polyethylene and a ¾-in. plywood substrate for the shingles. The poly will protect the ceiling's finish in case any hurricane-driven rain gets by the shingles. Finally, the crew removed construction smudges from the wood with orbital sanders, and finished it with two coats of penetrating oil.

Contrasting with the intricate living-room ceiling is the relatively simple post-and-beam treatment of the ceiling over the dining room and kitchen (photo facing page). This 4-in-12 gable rests on 4x10 rafters spaced 10 ft. apart. Atop the rafters are 4x4 purlins, which are flanked on each side by 1x2 ledgers. The ledgers do two things: they add a pair of reveals to each purlin, and they support the short sections of 1x4 T&G fir that compose the finished ceiling. Arranged in a random pattern with light sapwood interspersed among the darker heartwood, the T&G paneling creates a thatch-like texture that is right at home in a house surrounded by palm trees.

Compatible trim—Because it was his own house, Anderson had the luxury of deciding on details as the job went along. Instead of preparing extensive drawings ahead of time, he made frequent trips to the island, and kept in constant touch by phone with Carothers about the trim details that help to unify the house.

For durable floors that could stand up to sandy feet, Anderson used brick pavers called Versatile (Mutual Materials, Box 2009, Bellevue, Wash. 98009). These high-fire, 1-in. thick pavers are available in several colors, one of which closely matched the dark brown of the columns. Anderson used them on floors, shower walls, countertops and even as baseboards.

The ceilings were the first elements to receive a high level of finish, and their color and visual weight set the tone for the wood trim to come. Clear, dry Douglas fir became the material for doors, interior paneling, casings and cabinets. While the paneling and cabinets are made of ¾-in. stock, anything less than 1½ in. thick seemed just plain skimpy for casings and baseboards in the shadow of the substantial ceilings.

Mitered corners on casings and butt joints on door frames didn't seem to fit the style of the house, but the half-lap scarf joints used to join the beams in the kitchen ceiling looked right. Although Anderson used stepped joints for structural reasons in the ceiling beam, he began using them as well for exterior and interior door frames and for the casings surrounding them. Carothers and his crew applied variations of the detail throughout the house, including wood edging around tile counters, closet shelves and header moldings. To complement the stepped joint where door rails meet stiles, the crew mortised 2x2 baseboards into the casings.

Not surprisingly, this attention to detail had its price. At $125 a sq. ft., the house cost twice as much as anticipated, and its 20-month construction phase was also double the estimate. But Anderson is without regret. He introduced the crew to an integrated system of detailing, and as they investigated the possibilities within the system, the house became a superb collaboration between an architect and its builders. □

Purlins sit atop 4x10 rafters in the kitchen wing of the house. Note the scarf joint in the beam near the center of the photo. Details throughout the house emulate its angular lines. In the foreground, a teak table and chair custom made for Anderson in Thailand occupy the breakfast corner.

Adding a Craftsman Spa Room

Posts and beams raised on double-decker concrete slabs

by M. Scott Watkins

As a designer and builder of residential additions in the Washington, D. C., area, I find spa-room additions to be the most challenging. Not only do they require the usual reckoning of function, appearance and budget, but spas and spa equipment demand the use of structural systems and building techniques that are outside the realm of typical residential construction. As the third in a series of Craftsman-style spa rooms my company has designed and built, the Toftoy addition represents a refinement of our systems and techniques. Chuck and Patty Toftoy, like many couples, shopped the local real estate market and found that housing prices were sky-high. That convinced them to keep their old house and to gut most of the first floor, rebuilding it as a generous master-bedroom suite, including a master bath, sitting room and exercise room. A spa-room addition connected to the suite would be the focal point of the remodel. To minimize disruption to the Toftoys, we decided to build the spa room first.

Mechanical and moisture concerns—The Toftoys' house faces a quiet cul-de-sac and presents the modest look of a one-story suburban ranch house. The rear of the house originally offered a bland two-story brick façade facing a private, lushly landscaped backyard; the spa room is built off this façade (photo below). The exotic plantings and existing walks and patios in the backyard were a mixed blessing, helping to shape the design of the spa room, yet severely constraining construction activities (more on that later).

I began by sketching a simple, gable-roof addition measuring 9 ft. wide by 20 ft. long. The width was determined by the location of a stone walk, and the length by matching the roof pitch of the addition to that of the existing house and setting the ridge just below the sills of the existing second-story windows. With the basic outline established, I turned my attention to three major concerns. The first was where to place the spa equipment so it would be concealed yet conveniently accessible for maintenance. An easy solution to this problem would be to locate the equipment outside the spa room. However, we preferred to place the equipment inside for energy efficiency and easy access during bad weather.

Secondly, 300 gal. of 104° water has the potential to cause trouble. To forestall moisture damage, I needed to use durable materials and to provide proper ventilation and good drainage. Also, hot, humid air and chemical odors created by the spa needed to be isolated from the rest of the house or they would cause maintenance problems and discomfort. Third, because spa rooms are for relaxing and entertaining, we needed to select building materials that would engender a warm and intimate setting.

The solution to the spa-equipment problem was to suspend the floor over a crawl space and to install the equipment within the crawl space. Access is through a bench built with removable panels and positioned along the west wall (bottom photos, facing page). The 3-ft. high crawl space helps to isolate the equipment noise, and the bench provides the vertical clearance required for the spa filter and the air bubbler.

Both the crawl-space floor and the suspended floor are of concrete for durability and strength. The foundation walls are of insulated concrete block. The suspended floor, paved with quarry tile, adds thermal mass to the room for passive-solar heating. It also contains a radiant-floor heating system that's connected to the spa's circulation system.

Concrete stairs linking the spa room to the adjoining master suite serve as a barrier to prevent water from seeping into the interior of the house. French doors between the spa room and the master suite, complete with weatherstripping, contain the spa-room heat and humidity while connecting the spaces visually. This allows viewing from the spa of a fireplace in the sitting room. With the French doors open, the two spaces become one for entertaining.

I've always admired the work of Arts and Crafts-era designers, whom I consider to have been the masters of integrating structural and decorative elements of houses. In the tradition of Charles and Henry Greene and Gustav Stickley, I designed a projecting bay with a separate gable roof as the focal point of the room (top photo, facing page). On the exterior, the bay roof breaks up the massing of the façade into a more pleasing composition. Post-and-beam framing is exposed inside and out, and high quality Craftsman-style lighting fixtures (Arroyo Craftsman, 2080-B Central Ave., Duarte, Calif. 91010) provide an elegant accent to the exterior. Large expanses of glass lend an air of spaciousness to the small space.

Built over a suspended concrete slab, the Craftsman-style spa room links the master suite at the rear of the house with a carefully landscaped backyard. To make the room appear less top-heavy, redwood clapboard siding was installed with a 7-in. exposure for the bottom two courses and a 4-in. exposure for subsequent courses.

Photo this page: M. Scott Watkins

Site work—After much debate over how to protect the existing plantings during construction, we decided to build a temporary wooden ramp reaching from the driveway at the front of the house to the construction area at the rear. This ramp not only protected the plants, but aided us through all phases of construction, from pouring concrete to hauling materials.

The ramp consisted of seven 4-ft. wide by 16-ft. long panels linked together end to end and supported on simple 2x4 trestles spaced 8 ft. apart (photo next page). For safety, we attached 2x4 posts to the outside edge of the ramp and strung a nylon rope between the posts 36 in. above the ramp. Using duplex nails wherever practical, we were able to salvage much of the ramp material for another job.

A monolithic slab—The foundation had to support the perimeter loads of walls, floor and roof, while the crawl-space floor had to carry the point loads of the spa and its live load of water and occupants. I decided to meet all these requirements by pouring a 6-in. thick monolithic slab thickened to 8 in. at the perimeter (drawing next page). The slab rests on a 6-mil polyethylene vapor barrier and a 4-in. thick bed of gravel over undisturbed soil. Reinforcement consists of a 1-ft. grid of #4 rebar, doweled 6 in. into the existing foundation; we used a rotary hammer fitted with a 1-in. dia. carbide bit to bore the holes. To satisfy electrical-code requirements for potentially wet locations, we turned up one piece of rebar so that it would protrude above the top surface of the slab. This allowed the electrician to ground the spa equipment to the rebar.

Souping up the concrete—Our construction area was inaccessible to concrete trucks, so we had to consider other options for concrete placement. The use of wheelbarrows was out of the question because the ramp was too steep. Pumping the concrete over the house sounded good, but the monolithic slab and suspended concrete floor required two separate pours, and the cost of hiring a pump truck for more than one delivery was prohibitive. We decided instead to attach 2x10s to the top of our ramp to form a concrete chute.

The key to this operation, though, was Eucon 37, a superplasticizing concrete admixture available from the Euclid Chemical Co. (19218 Redwood Rd., Cleveland, Ohio 44110). This product is classified as a non-chloride (non-corrosive), high range (high slump), water-reducing superplasticizer, and conforms to the ASTM C494 specification for concrete admixtures. It's designed to be added to relatively stiff (2-in. to 4-in. slump) concrete to in-

Inspired by the work of the Greene brothers and Gustav Stickley, the spa room features mortise-and-tenon framing, finished with clear oil, and has a shallow bay as its focal point (photo above). Posts are spaced to accommodate standard-size awning and casement windows.

A redwood bench inside the room appears to be exclusively for sitting (photo far right). But the bench also provides vertical clearance for the spa equipment located beneath the floor, and offers access to the equipment through an opening that's created by sliding off a pair of removable panels (photo right).

From *Fine Homebuilding* magazine (February 1990) 58:58-61

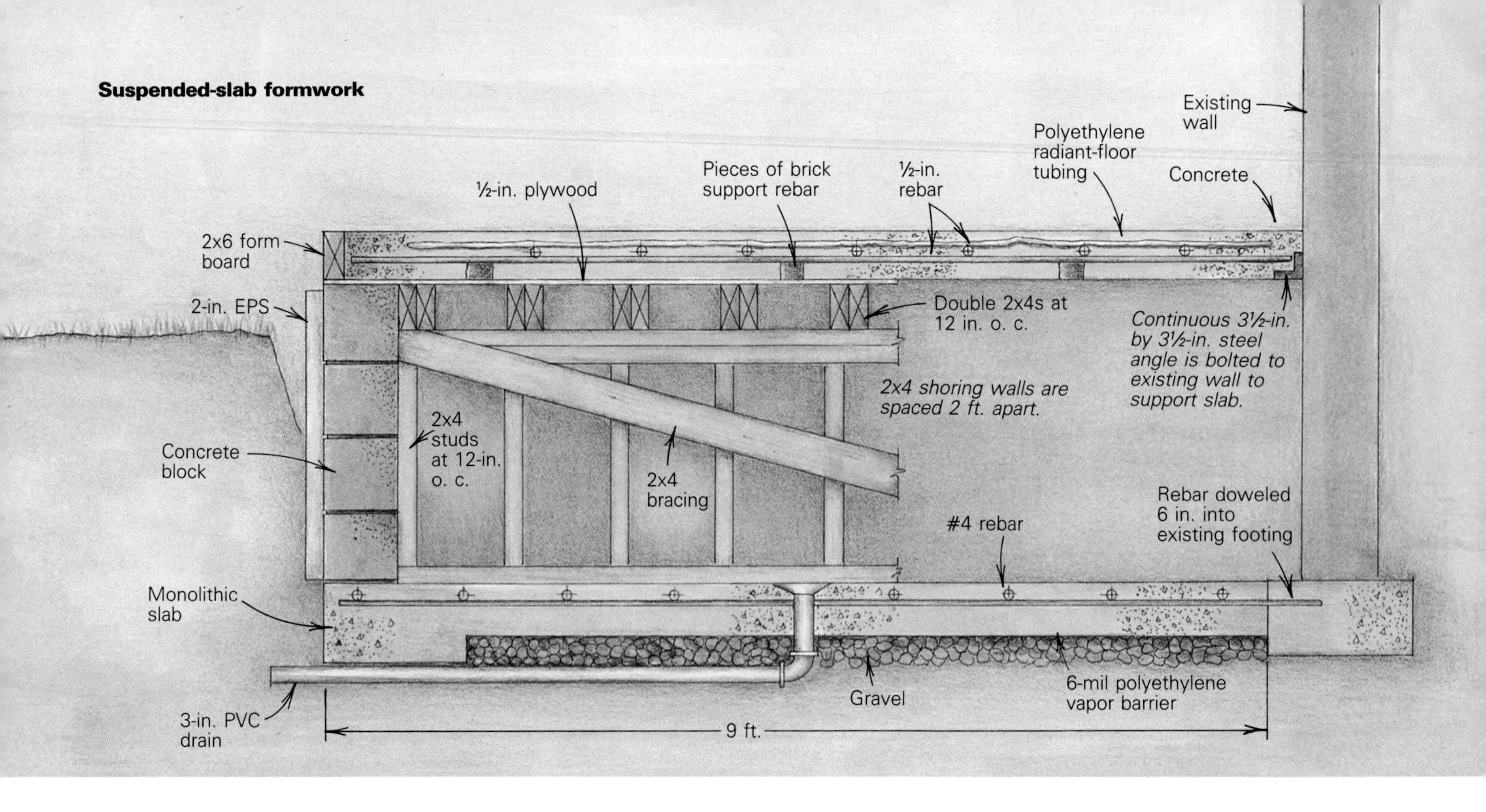

Surrounded by exotic plantings in the backyard, the spa room was inaccessible to concrete trucks, and the cost of hiring a concrete pump was prohibitive. Instead, the concrete mix was modified with a superplasticizing admixture and a double dose of air-entraining admixture, then chuted down to the forms over a plywood ramp. In the photo above, concrete is being placed in the forms (drawing above) for the suspended main floor. The worker is standing on a block-out form nailed to the top of the forms to keep concrete out of the spa cavity. A second block-out form to the rear corresponds to the location of a redwood bench that will open into the crawl space. Snaking through the slab is polyethylene radiant-floor tubing, tied to a one-foot grid of ½-in. rebar with copper wires.

Photo: M. Scott Watkins

crease its fluidity without adding excessive amounts of water, which would reduce the concrete's ultimate strength. The product added about $35 to the cost of the concrete for both pours.

With the help of our local concrete batch plant, we designed a customized mix to use with the superplasticizer. The plant delivered a six-bag, 4-in. slump concrete with double the normal amount of air-entrainment admixture. Air entrainment creates microscopic air bubbles in concrete and is usually used for protection against the freeze-thaw cycle. In this case, doubling the admixture helped to lubricate the aggregates in the concrete for greater mix fluidity.

When the concrete truck arrived, we added the Eucon 37 at the rate of 12 oz. per 100 lb. of cement. The truck operator then agitated the mix in the drum at its highest speed for five minutes.

This was our first experience with using a superplasticizer. Prepared for the worst, we were pleased to see gravity pull the concrete down the chute the first 90 feet. Only in the last section of the chute did we need to help the concrete along with shovels.

Another benefit of using a superplasticizer was the ease of finishing the concrete. Because the admixture reduces the water/cement ratio, we could screed off the concrete and immediately begin floating it. There was no waiting for excess water to bleed off. After floating, we covered the slab with polyethylene for three days of undisturbed curing. We also wet the slab and kept it covered between work periods for about three weeks.

A suspended radiant floor—With the concrete-block foundation walls in place, we carefully planned the construction and alignment of the forms for the 6-in. thick suspended concrete floor. Any mistakes here would be cast in concrete if the forms weren't accurately aligned to accommodate the tile floor, the shallow bay and the post-and-beam framing.

First we ordered a custom ceramic tile/fiberglass-reinforced spa from a local retailer. The spa manufacturer, Sunrise Industries (1463 E. 223rd St., Carson, Calif. 90745), assured us by telephone that although our "Sapphire" model was a custom order, they would build it with a stock mold and thus we could rely on the dimensions provided in their sales literature.

We formed the suspended floor by building a temporary deck flush with the top of the block wall and installing a 2x6 rim around the perimeter to contain the concrete (photo and drawing, facing page). The deck consisted of ½-in. plywood over doubled 2x4 joists running the length of the spa room, 12-in. o. c. Shoring beneath the deck consisted of diagonally-braced 2x4 walls 2 ft. o. c. that rested on the crawl-space slab. We used duplex nails wherever possible so the formwork would be easy to remove later.

We left a manhole in the decking in the area of the spa-room bench to provide access to the crawl space for stripping the formwork. We marked a reference line on the back wall of the house to correspond to the top of the suspended floor and nailed down simple 2x block-out forms at the spa and bench locations to keep concrete out of those areas. The spa form was slightly lower than the perimeter forms to create a slope of 1/16 in. per foot in the floor for drainage into the spa. With the formwork in place, we now had a safe working platform from which we demolished portions of the existing walls.

With the formwork complete, we wired together another 1-ft. grid of #4 rebar and propped it above the deck on pieces of brick. To support the back of the slab, we lag-bolted a continuous length of 3½-in. by 3½-in. angle iron to the house, spacing the bolts 2 ft. o. c.

Now it was time to install polyethylene tubing for the radiant floor. The tubing, accessories and layout design were provided by the Wirsbo Company (7646 215th St. W., Lakeville, Minn. 55044). Wirsbo tubing is manufactured in Sweden and is specifically designed for radiant-floor heating. We fastened the tubing to the rebar with copper twist ties provided by the manufacturer, forming loops one foot apart. Where tight bends were required, we attached special angle supports (also available from the manufacturer) to prevent kinking. The tubing runs in one continuous length, starting and ending in the crawl space.

Finally we were ready to place the concrete, but decided to wait until the spa arrived. It's a good thing we did. Despite the manufacturer's assurances, the spa varied from the published dimensions just enough so that we had to completely rebuild the forms for the bay and the block-out form for the spa. Only after checking and rechecking the dimensions of the rebuilt forms did we order the concrete.

With the aid of the ramp, chute and superplasticizer, we placed the concrete without difficulty (photo facing page). Once we screeded and floated the slab, we covered the floor with two layers of 6-mil polyethylene for 28 days of curing, then stripped the forms and shoring, removed the 2x10 concrete chute from the ramp and slid the 1,100-lb. spa down the ramp to the construction area. With the spa lowered into its final position in a bed of grout, we built a temporary plywood cover for it to offer protection during the remaining construction.

Post-and-beam framing—After fussing with concrete for over a month, it was a relief to start framing. Select structural Douglas fir is strong, looks good and is easy to engineer from published span tables, so we used it for all the exposed framing members.

The walls consist of posts and beams with 2x4 stick framing between the posts. The bottoms of the posts are notched to straddle 2x4 pressure-treated sills, and the tops are mortised into 6x top plates. The gable-end top plates are notched into the the side-wall plates at the corners so that they bear on the corner posts. Intersecting plates at the bay are lap-jointed together. Mortise-and-tenon joints are wedged tight and all connections are spiked together with 20d ring-shanked stainless-steel nails (Swan Secure Products, 1701 Parkman Ave., Baltimore, Md. 21230). Before raising the roof, we routed a groove along the tops of the top plates and called in the electrician to wire the hanging lanterns.

The roof structure consists of a 6x6 ridge beam and purlins, 3x4 rafters and 3x6 barge rafters supporting T&G pine decking. The outboard ends of the ridge and purlins rest on built-up 4x4 posts concealed in the gable end; the inboard ends rest in pockets chiseled out of the existing brick wall. The inboard ends were further supported by 1½-in. by 1½-in. by 2-ft. long steel angles bolted into the brick mortar joints.

Once the roof deck was nailed down, we installed 30-lb. felt, 3-in. thick Thermasote rigid-foam insulation (Homasote Company, P. O. Box 7240, West Trenton, N. J. 08628), a second layer of felt, and aluminum shake roofing (Reynolds Metals Co., Reynolds Rd., Ashville, Ohio 43103). This roofing is priced between asphalt shingles and wood shingles, and will outlast asphalt. I would have preferred wood shakes or shingles, but the Toftoys wanted the new roof to match the existing aluminum roofing on the house. The exterior bevel siding is clear all-heart redwood clapboard siding. To balance the appearance of the addition, which seemed top-heavy, we used a 7-in. exposure for the bottom two courses and a 4-in. exposure for the rest.

A hybrid heating system—Once the spa equipment was installed, Elwood Huff, our heating subcontractor, went to work. He spliced the radiant-heating system into the spa system on the supply side of the spa water heater. The radiant heat is controlled by a standard thermostat that is mounted on the wall. When it switches on, both the radiant-floor heating pump and the spa pump (which normally cycles on its own separate 24-hour timer) are activated. That's because the water flow produced by the radiant-floor pump alone is insufficient to activate the flow switch built into the water heater. A low-voltage solenoid valve prevents water from entering the radiant-floor heating system when only the spa pumps are on. At the return end of the radiant-floor loop, a check valve prevents reverse flow.

On the coldest days the floor system cycles on and off about every 10 minutes during the night and on overcast days. I think this short cycling is the result of the thin spa cover allowing too much heat to escape from the spa water and directly heat the air. The Toftoys plan to try a better spa cover in the future to allow the radiant-floor heat alone to heat the room. □

M. Scott Watkins is a designer and builder in Arlington, Virginia.

Drawing: Michael Mandarano

Budget Bungalow

Behind fanciful porch columns and carved rafter tails sits a small energy-efficient house

by Gerry Copeland

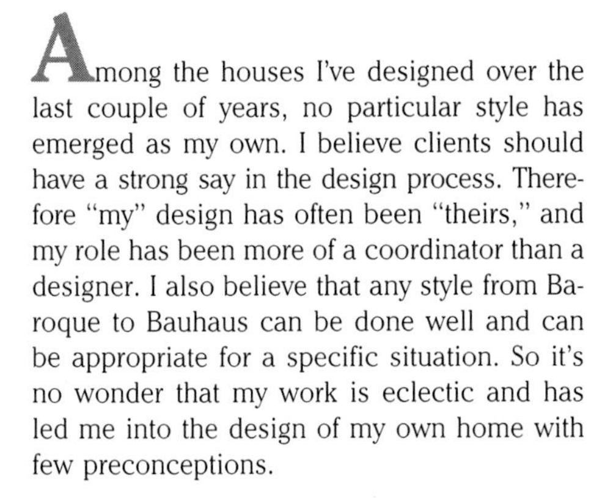

Among the houses I've designed over the last couple of years, no particular style has emerged as my own. I believe clients should have a strong say in the design process. Therefore "my" design has often been "theirs," and my role has been more of a coordinator than a designer. I also believe that any style from Baroque to Bauhaus can be done well and can be appropriate for a specific situation. So it's no wonder that my work is eclectic and has led me into the design of my own home with few preconceptions.

Influences—After looking around the older, established neighborhoods of Spokane, and out in the suburbs, I knew that I didn't want to live in a homogeneous neighborhood where all the houses look alike—the 1982 Street of Dreams, or the 1975 Solar Home Show, or the current trendy houses with many hips and arch-top windows that can supposedly be plunked down anywhere from Spokane to Boston.

Having moved to Spokane from the mountains of northeastern Washington, I wanted some elbow room—a large or double lot. I found what I was looking for on the south hill, a mixed neighborhood of 1910 farmhouses, Craftsman-era bungalows, 1960's flat-roof moderns and a few infill houses of widely varying quality. Because of this broad mix and the corresponding range of value, the price for this 130-ft. by 150-ft. lot was only $16,000. It's bordered on two sides by rugged, unbuildable land with huge rock outcroppings and mature pines.

Down the block there are a couple of real gems—well-maintained Craftsman-style houses with exquisite eave and bracket detailing, window and door trim, and nicely proportioned columns and capitals. They were built during the early years of this century, Spokane's heyday of home building. At that time, Spokane was an important financial center and railhead for gold and silver mining in the northwest. Many of the homes on the south hill were built then; I decided to design a home that would be a compatible neighbor. Not to be a copyist, I spiced up the design with details from temples and villages I had seen while living in India as a Peace Corps volunteer and while traveling in Nepal and Sri Lanka.

Besides the stylistic influences on the design, there were budget constraints and energy efficiency to deal with. I controlled the cost in part by doing some work myself and by using my construction company without the usual overhead and profit markups. The house encompasses 1,750 sq. ft. of heated living space, and my total cost was $75,000 (not including the lot).

This house was one of the first in Spokane to be built under the super energy-efficient building codes. It has R-30 walls, R-50 ceilings, R-19 basement walls, low-e double-glazed wood windows, insulated steel doors and a whole-house air-to-air heat exchanger/electric furnace.

These high levels of insulation led to an interesting juxtaposition of turn-of-the-century detailing and state-of-the-art energy efficiency. For instance, the exposed rafter tails are actually the top chords of parallel-chord trusses. Two-by-four chords would have been sufficient structurally, but I ordered 2x6 top chords so I could carve them to the design I wanted.

A complex roof made easy—The dominant design features of the house are the distinctive double-pitched roof and the detailing at the eaves, gable ends and porch. The roofline with the steep 12-in-12 pitch over the second floor and the 7-in-12 pitch over the porch and first floor is Japanese in origin. Because of the thick space needed for insulation and air space over sloped ceilings, I used a variety of parallel-chord trusses, the most complex being a combination of two trusses: one with a 12-in-12 slope and the other with the 7-in-12 slope (drawing, p. 126).

I am always amazed when I show my plans to a truss designer and he says, "I can do it." Not only can truss manufacturers do the typi-

From *Fine Homebuilding* magazine (Spring 1990) 59:52-56

cal gable configurations, but they can do hip assemblies of different lengths with compound cuts, too (drawing, p. 127). Basically, any roof that can be framed conventionally (stick-built) can be prefabricated with trusses. I keep comparing the costs of stick-built framing—using wood I-beams—to prefabricated trusses, and so far the trusses always come in lower. One of the most satisfying days in the construction of the house was when this complex, prefabricated roof went together smoothly and quickly. Despite the versatility of manufactured trusses, there were a few areas of the roof that we had to stick-build.

I used the parallel-chord trusses in order to get the depth for enough insulation. R-38 is the code minimum in this area, but I needed R-50 to compensate for all the windows in the living area. This takes approximately 18 in. of insulation, be it fiberglass batt or blown-in cellulose. Against my better judgment, I used a combination of R-30 and R-19 batts and tried to use blown-in cellulose to fill the voids where the truss configuration prevented the batts from snuggling up against each other.

I would have preferred blown-in cellulose for the whole ceiling because this method fills in and around all the truss members. But the local building inspectors won't allow this because they think that the blown-in insulation will settle and leave gaps. My experience has been that this doesn't happen, and if it did, it would be easy to add more at the top. With air gaps at their sides, fiberglass batts lose approximately 50% of their insulation value. The best system to use might have been the Blow-In-Blanket System (BIBS) (Ark-Seal International, Inc., 2185 S. Jason, Denver, Colo. 80223) where loose insulation materials are mixed with an adhesive binder that prevents settling and provides a tight fit in and around framing members.

By unifying such diverse architectural influences as the farmhouse across the street and the temples of Nepal, architect/builder Copeland also managed to build 1,750 sq. ft. of living space for $75,000 (top photo). The carved porch columns and capitals (photo above) were bandsawn from glulam beams. The curvacious rafter tails, cut with a jigsaw and a router, are actually the 2x6 top chord of the parallel-chord trusses.

Porch details—The front porch, with its carved columns and capitals in deep red and its blue ceiling (photo left), was inspired by Nepalese temples. Because I needed a thick piece of wood that wouldn't check or twist, I used 5⅛-in. by 6-in. glulam beams for the columns. I cut the decorative details on all four sides of the columns with an old 20-in. bandsaw. By using a ¼-in. blade I was able to cut the small-radius curves without much difficulty. With a sharp blade, patience, and a temporary support to hold the other end of the glulam, I cut the 6-ft. long columns myself. I trued up the curves with a 4-in. belt sander and then eased all edges with a ¼-in. roundover router bit.

I made the capitals the same way, and then screwed them from the top into the columns with ⅜-in. by 12-in. lag bolts. The col-

The great room dominates the interior of the house, serving as living room, dining room, kitchen and staging area for family activities. In a whimsical spirit, Copeland installed an upside-down capital (like those used on the front porch) to distribute a point load from above to the glulam beam.

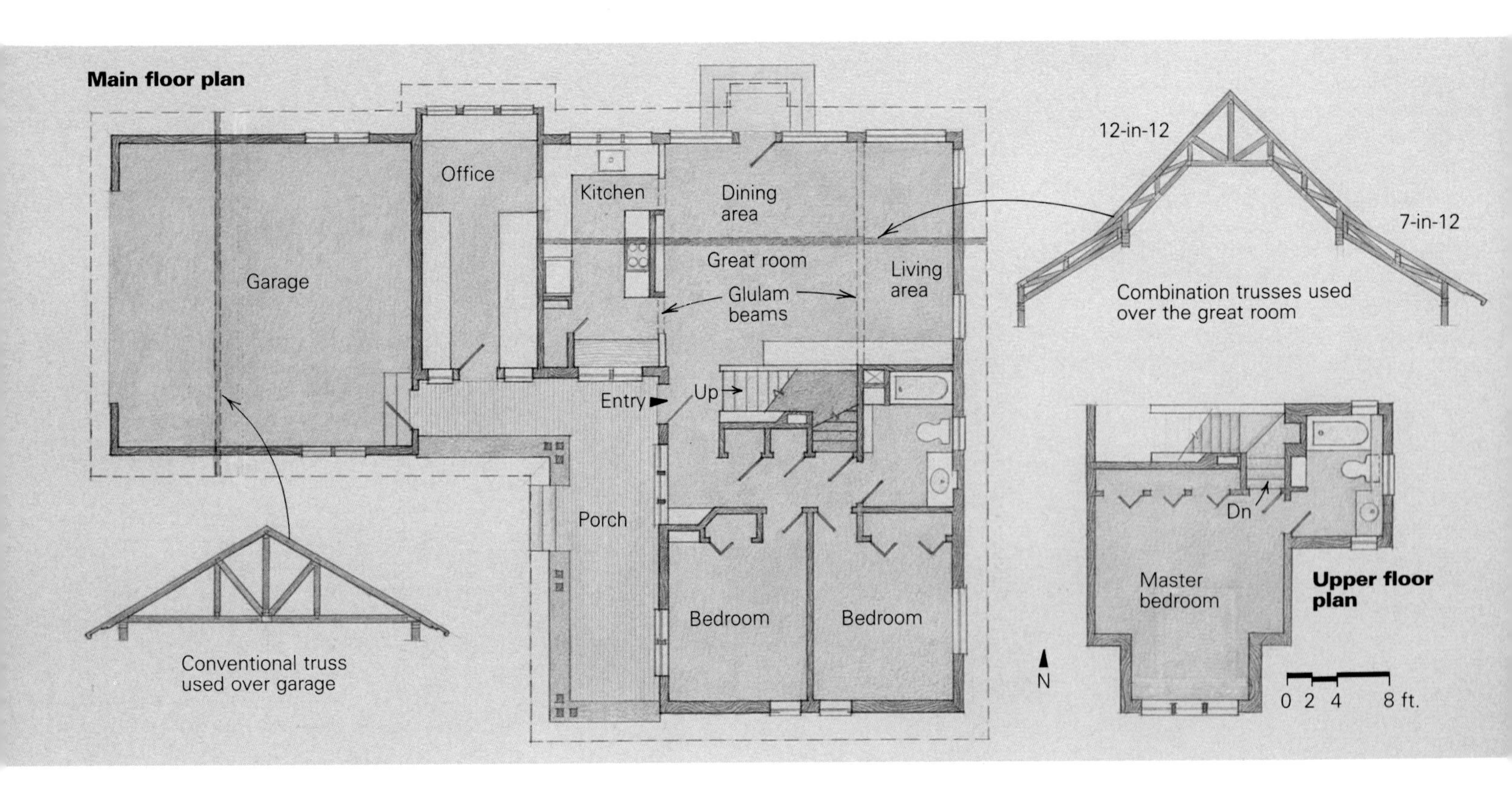

umn bottoms are bolted through decorative steel bases, which I had custom made.

The glulam columns and capitals were filled and sanded, primed with an oil-base primer and then finished with several coats of exterior latex. The blue porch ceiling is ordinary drywall primed and painted with a color I call "infinite blue" because it doesn't allow depth to be easily perceived. Depending on the lighting conditions, the ceiling may appear a few feet away or stretch off into infinity.

Wooden porch railings are traditionally a maintenance problem. They suffer the ravages of the weather, the procrastination of the homeowner, and the weight of people sitting on them. I wanted to circumvent these problems, so instead of a railing, I decided to build a wall that people could sit on. I wanted this wall to be impervious to water from seeping flower pots and milk from the kids' spilled cereal bowls, so I cast 3½-in. thick by 14-in. wide "seats" right on top of the concrete footing walls around the porch. At the top edges I nailed standard 1x1 cove trim inside the forms to create rounded edges (top photo, p. 125).

The finishing of these walls has become a minor point of contention between my wife and me. Originally we planned to apply a sack finish to the seat and wall, which involves rubbing the cured concrete with mortar to hide form lines and air pockets. My wife hates the way the walls look and is anxious for me to finish them. But lately I've grown to like the honest, unpretentious look of the raw concrete...or could this be the procrastination of a homeowner?

Carved rafter tails—The carved rafter tails and narrow fascia details were inspired by a house nearby built in the early 1900s. The house had always looked Oriental to me, and sure enough, I learned that it had been occupied by a Japanese-American family. I drew up a close copy of its rafter tails, and then made a full-size template out of matt board. We used this to draw the outline onto each rafter tail as we got to it.

We stacked and unstacked the different size trusses many times in order to cut each one with an orbital-action jigsaw (the best quality sharp blades were a must). In order to show off each of the 151 meticulously cut tails I used a narrow fascia consisting of a 2x2 sub-fascia with a 1x2 on top of it to cover the exposed edge of the roof sheathing.

Over the exposed rafter tails we ran ⅝-in. T-111 plywood with the groove side down, exposed below. It simulates V-groove 1x4 soffit material, but costs less and offers greater structural strength. A 4-ft. width of this plywood runs around the roof perimeters, and the rest of the sheathing above is standard ⅝-in. CDX plywood. Finally, the rafter tails and the underside of the T-111, along with the frieze board at the top of the exterior walls and the fascia, were painted a deep red.

The farmhouse across the street from us has decorative trusses on the gables, so I chose a similar treatment for our house. The barge rafters are supported by curved brackets, made from glulams and lag-bolted into the wall framing. The collar ties and king posts are standard framing lumber.

Because we live in a forested area, I was concerned about fire and didn't want to use a wood roofing material, so we installed laminated fiberglass shingles (Prestique Plus, Elk Corp., 14643 Dallas Pkwy., Suite 1000, Dallas, Tex. 75240). They come with a 40-year warranty, and the "weather wood" color we chose is almost exactly like weathered cedar shakes.

Patterned siding—To add interest to the beveled cedar siding, we ran two courses of 8-in. boards with the rough side out, alternating with one course of 4-in. boards, smooth side out. Although I hadn't seen this done with clapboards, it was a common pattern on

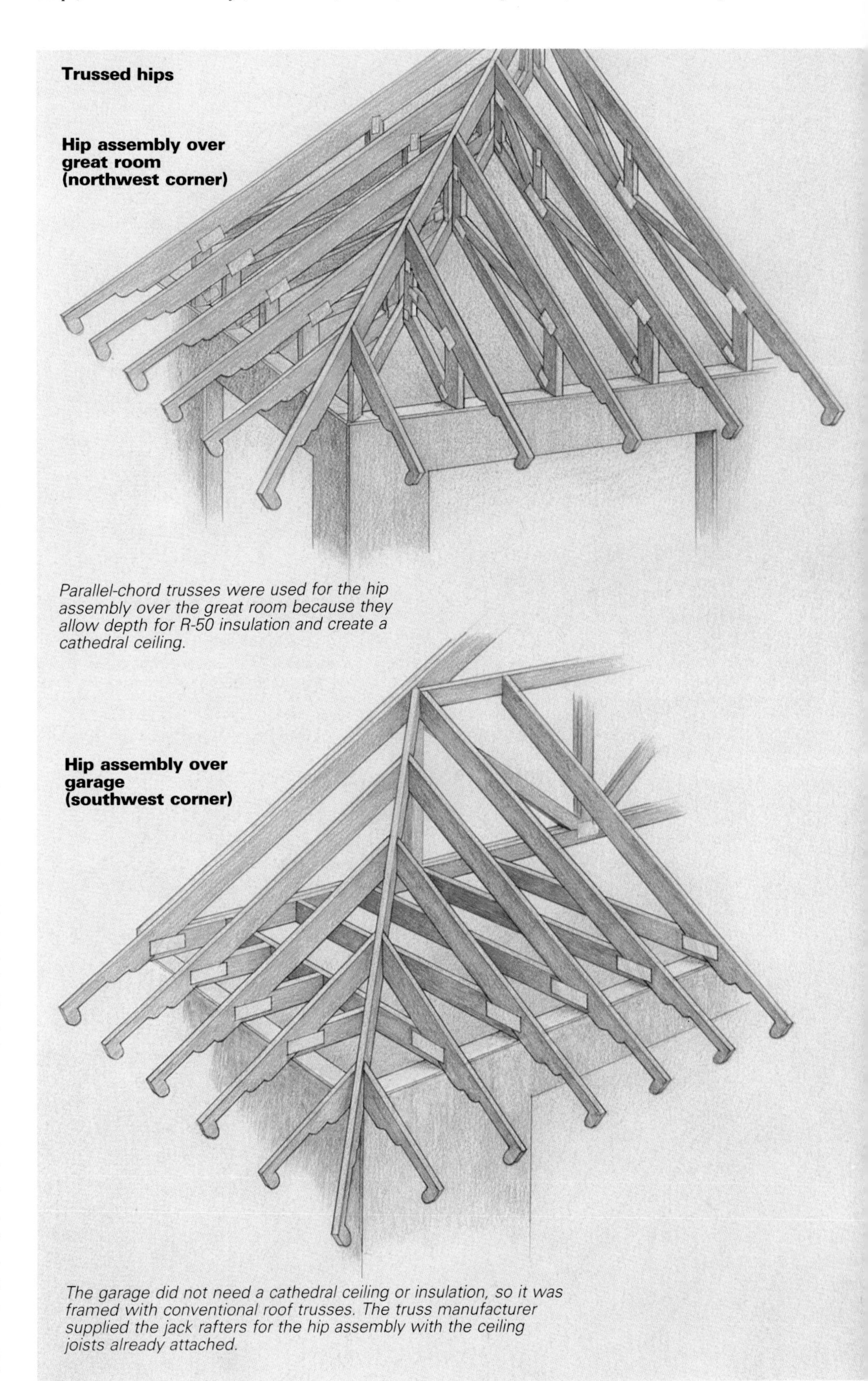

Parallel-chord trusses were used for the hip assembly over the great room because they allow depth for R-50 insulation and create a cathedral ceiling.

The garage did not need a cathedral ceiling or insulation, so it was framed with conventional roof trusses. The truss manufacturer supplied the jack rafters for the hip assembly with the ceiling joists already attached.

Drawings: Gary Williamson

shingled houses. This siding was factory stained to a light grey/brown color that's supposed to look natural over time.

We used a story pole to install the siding, laid out so that along the eave walls the bottom board and the top board were of equal widths. The real trick in keeping the imperfect clapboards straight and horizontal was using a siding gauge—just a short 1x2 with a 1x2 crossiece screwed to it. The carpenters ran it along under each course when nailing to gauge its distance from the board below.

We beveled the butt joints at 30° using a Delta Sawbuck. Despite all protestations to the contrary, wood does shrink in length, and besides, a beveled joint is more forgiving. Unfortunately, even though we nailed with 7d galvanized siding nails through the sheathing into the studs behind, some of the 8-in. clapboards (standard #3 tight-knot quality) still moved and wriggled enough to make a nearly perfect job look to be just normal.

To make the windows and doors stand out from the natural cedar siding, we surrounded the standard brick-mold window trim with cedar 1x4s, smooth side out. This was thick enough for the beveled cedar siding to butt up to. We painted this trim, along with the doors and windows, the same deep red.

Because the backyard is such a private space, Copeland saved most of the windows for the rear wall of the house. He also manipulated the eave line to create protective roof overhangs for the back door and the projecting window bay.

Salvaged from the staves of 70-year-old acid vats, the vertical-grain fir used for the kitchen cabinets doesn't look too bad after resawing and lacquering.

Upside-down capitals, recycled fir—The interior of the house is dominated by a great room that serves as living room, dining room and kitchen (floorplans, p. 126). With a wall of windows and a true divided-light door, I oriented this room toward the backyard (top photo), which is very private and will eventually become a Japanese garden. The windows and door are Oregon Ponderosa pine, cased with clear hemlock. The ceiling is 1x6, T&G cedar (photo, p. 126). The cedar had been in storage outside under black plastic for six years. We rejuvenated it by running the exposed side through a thickness planer. The ceiling boards were scribed to the walls along the sides and at the ends so that no trim was needed and a cleaner look was achieved.

Two glulam beams pierce the ceiling of the great room, countering the outward thrust of the roof on the rear wall and providing bearing via the short post and upside-down capitals for the upper ceiling/roof. I used the same capital detail that I had used on the front porch—this time, upside-down to distribute a point load from above to the glulam beam below. I finished the interior glulams by sanding off any blemishes, cleaning with a damp rag and finishing with clear polyurethane.

The kitchen cabinets were built out of a vertical-grain fir salvaged from 70-year-old acid-tank staves. My cabinetmaker friend, P. D. Voelker, got a deal by taking apart some large acid tanks from a nearby mine that was being dismantled. From the 3-in. by 6-in. vat staves, he cut the 1x3s and 1x6s that he used to make face frames, doors and drawers.

Vertical-grain fir looks similar to *sugi*, or Japanese cedar, a traditional wood used to trim Japanese homes. The fir that we used, taken from old-growth, 6-ft. dia. trees, has aged over the last 60 years or so and looks great now recycled into cabinets (bottom photo).

The finish on the cabinets is several coats of clear lacquer. The natural finish on the fir glulams, the fir cabinets, the hemlock casing and the clear, non-yellowing sealer (Benite by Daly's Wood Finishing Products, 3525 Stone Way N., Seattle, Wash. 98103) used on the cedar ceiling yielded a surprisingly uniform bunch of colors.

Building a new house in an existing neighborhood often invites resentment for upsetting the status quo and filling up another small piece of wild land. The reactions from this neighborhood have been positive, and when a local drives by and shouts, "Nice house," I feel good about being here. □

Gerry Copeland is an architect and builder in Spokane, Washington.

Handcrafted in Stone and Wood

An architect turned craftsman builds his house of fir, cherry, mountain stone and slate

by Philip S. Sollman

I was born in the eastern part of the Netherlands, a region of many rural farms and medieval estates. I've always admired the rich detail and intimate spaces of its architecture, so it was natural for me to seek an education in that field. After getting my degree in architecture from Penn State, I worked for an architect who specialized in residential design. Soon, a friend and I formed a small construction business. We designed and built many houses, but I wanted to spend more time on the details that contributed to a beautiful house. I finally left the business and went to work as a craftsman—mainly in wood—so I could work more freely and sculpturally.

My wife Jeanne is a ceramic sculptor. A few years ago we bought 10 acres in Pennsylvania, where we planned to build our house and workshops. We knew the task would take several years, so we made it the focus of all our resources and energy. During the first year, our time was divided between earning a living and building our house. Jeanne's aptitude for stonework was especially helpful to me. Later, as the demand for her sculptural work increased, I was able to spend more time on the house and less on my own commissions.

The design—I've always liked the houses of Greene and Greene and Frank Lloyd Wright because of the beautiful way they integrate craftsmanship and structure and the thoughtful manner in which they use natural materials. I don't like stark, International-style houses without enriching detail. I wasn't interested in copying anyone's style, but I knew

Philip Sollman is a self-taught craftsman working primarily in wood and stained glass.

The house (top) is set in a southern slope on a 10-acre plot near State College, Pa. Although the emphasis is on craft, not engineering, the south-facing glass and thermal mass keep the house warm most of the winter. A woodstove and fireplace burn only 1½ cords a year. Sollman framed the house with heavy fir timbers, decorated with a step motif (above). Variations on the step recur throughout the house.

I wanted to use stone and heavy timbers for the warmth and security they give. I was looking forward to working these materials into forms that suited them, and that we liked.

The basic plan of the house has changed very little since I made my original sketches. I set the house and future workshop into the southern slope, taking advantage of the sun and some earth sheltering to the north and west. The approach to the house is from the high north side and the entry door is between the two floors, under an overhang that extends from the gabled roof. A path meanders down to the entry flanked by two heavily planted earth berms. My idea was to create a somewhat dark and intimate entry that would open into light and spaciousness as you descend into the living room.

I wanted to heat with the sun, but I didn't want the house to be strictly an engineering job. I rejected the idea of a Trombe wall right away, because I didn't want to block the view to the south. This decision made me feel that I'd have to install a backup heating system and every energy-saving device I could think of. I built many of these into the foundation. I laid copper pipes in the slab for radiant heat, and I installed ducts under the floor through which warm air is drawn from the ceiling and blown out through grates in the floor. I also built in an outside air source for the fireplace, and a hookup for a woodstove.

One of my major efforts was to build in a 4-ft. wide by 7-ft. high tunnel that would eventually be 80 ft. long and link the house to the future workshop. This would give us room for food storage, and would let us get to the other building without going outside. It might also help cool the house in the summer because its temperature is always between 55°F and 60°F.

The house is just under 2,000 sq. ft. All the major spaces are either along the south wall or run the full width of the plan. The two north-south interior bedroom partitions slide into pockets. This opens the bedrooms up to the two-story living area below, and improves air circulation in the summer.

We've been through two heating seasons now, and have never used the radiant system. The house stays warm all winter without it. We don't even need the woodstove much. We burn only about 1½ cords of wood per year.

Although we didn't begin construction until

Photos: Jeanne L. Stevens-Sollman

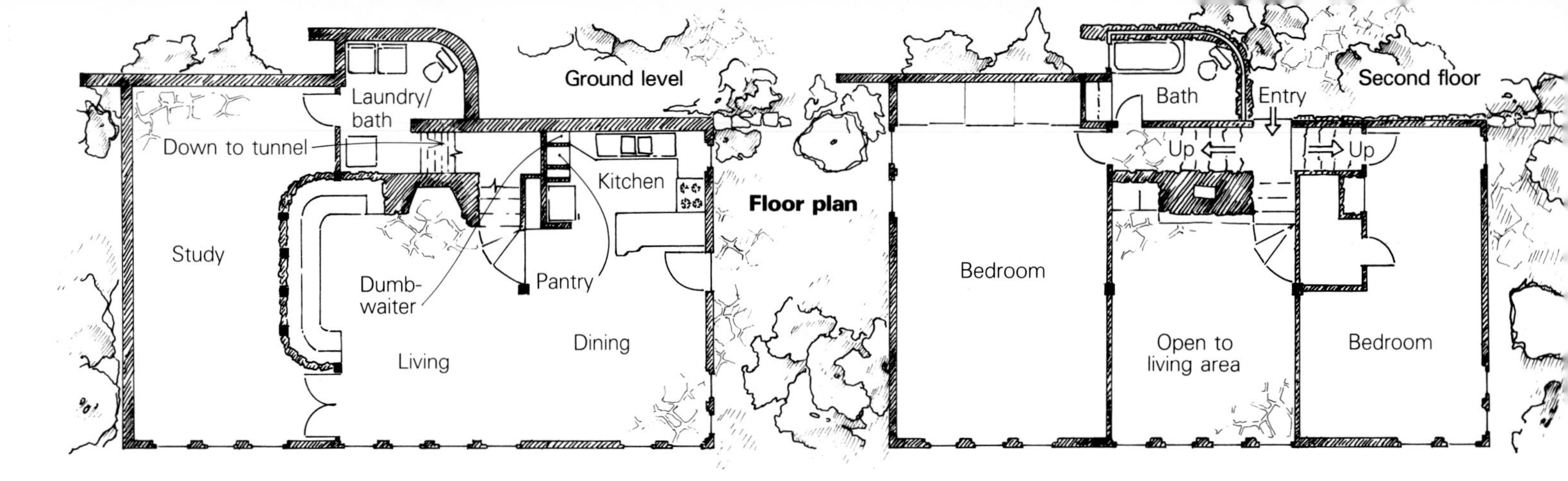

June 1979, I built the walnut entry door and some lamps beforehand with the house in mind, to get a feeling for the kind of workmanship I'd be doing once we really got underway. I had originally intended to trim the interior out in walnut, but I changed my mind when I saw how it contrasted with the fir structural timbers. Cherry goes with the color of the exposed structure beautifully, so I used it for the trim, paneling and cabinetry in the rest of the house.

When we finally did break ground, I worked on the foundation and fireplace for about a year, the framing for another year, then finally made windows, doors, lamps, cabinets and built-ins, door knobs, and switchplate and outlet covers.

Stonework—Although fireplaces are not efficient heaters, I think they anchor a house to its site and give a space a feeling that a woodstove can't. We love our fireplace (photo facing page, top) and use it until December, when we hook up the woodstove, which stays in the hearth until early spring. I did try to make the fireplace as efficient as possible. It has a Rumford-style firebox—a shallow, tall opening with sharply angled side walls to encourage radiation (precisely proportioned dimensions are based on the width of the opening). I added an outside air source and hot-air circulation through ducts in the chimney and floor—a sort of home-fashioned Heatilator.

Building the fireplace and chimney was the first time I'd ever laid stone. My vision was of neatly dressed stones laid up in a tight horizontal pattern, but we used what's locally known as mountain stone, and discovered early on that you have to find the right shapes rather than make them. It was a matter of training the eye. I laid the corners with the squarest and largest stones in an overlapping bond, then filled in with the irregular shapes. There were many more irregular shapes than square ones, and what I had envisioned is not what I ended up with. Happily, I like the end result better. The material had educated me in its use.

I designed the steps from the entry into the living room to suggest a cascade, to make the levels appear to flow together. We bought the irregular flagstones from a nearby quarry. As we expected, our skills at fitting the undressed edges together improved as we progressed. Ironically, this raw material is actually cheaper than vinyl flooring that's made to look like stone. The bargain ends when you get to installation, though.

We worked on the floor during the winter of 1980, after the house was closed in but before it was insulated. We used a salamander heater—a kerosene heater with an electric fan—while we worked, and at night after we left, the temperature remained above freezing. The solar orientation and thermal mass were already proving their effectiveness.

Working with wood—I began framing in the winter of 1979-80. I wanted the structure, much of which would be visible, to look sculpted, so I decided to decorate the timbers with a step motif (small photo, previous page) as they were being erected. I had to shape and sand each beam before I put it up, and I could only finish one or two of them a day. This made the raising slow, but it gave me time to think over my next move: how to relate the inside of the house to the outside, the windows to the doors to the siding and interior trim. This is when I began to develop many of the ideas for the details and decoration.

I wanted the form of my work to flow with the patterns and directions of the wood. Just as the fir had suggested the use of cherry, the cherry's natural patterns eventually suggested new forms. As I continued to work on the structure and its details, the house in turn was shaping my direction.

I wanted the siding near doors and windows to look as though it was reacting to the placement of the openings—like the rippling of a pond when you drop a leaf in the water. Of course, this meant a lot of work, not only custom trimming all the siding to fit around the door and windows, but also building the doors and windows themselves.

I built the window frames out of dry native pine as either single or multiple units. They were sealed, primed and painted before installation. I took care to select the clearest pieces of pine for each sash. If there was a knot, it had to be near the outside edge, away from where the glass was set in. Otherwise the risk of cracking a pane would have increased, because of the way wood shrinks and swells around a knot, exposing it to apply pressure against the glass.

All of the finished sashes were sent to a glazier, who fitted each window unit with double-insulated glass. The sashes were set into the already installed frames, and sealed with weatherstripping tape. Most of the operable windows are hinged at the top. They open out at the bottom, protecting the inside sill from rain. They don't always have to be closed during a shower. After nearly three years, they still fit tightly and open easily.

The inside woodwork began at the baseboard. Along the south wall it ran nearly 30 ft., and had to be spliced in four places along its length and then fitted to the wall. To make each long, diagonal splice as subtle as possible, I sorted through my boards to find color and grain patterns that matched. Unfortunately, I'd laid the floor right up to the wall, so I had to scribe-fit the bottom of each baseboard to the stone floor. I should have laid the stone just short of the wall and grouted it in after each baseboard was set.

Baseboards, jambs and window trim stand in relief from the paneled walls. It's all cherry. The horizontal paneling is tongue-and-groove all around. Its ends are joined into the door and window trim to eliminate the need to plug screw and nail holes. These would have shown up as blemishes on the rich figure of the paneling. Installing it this way isn't really hard; it just requires three hands.

Switchplate and outlet covers—The stepping detail from the fir framing members popped up in the baseboard and window trim, too. Rather than downplaying the electrical outlets, I decided to make them a major element in the composition of the wall, since our hands and eyes are always searching them out. It occurred to me to step the trim around the outlets, letting it pick them up wherever they were. This made focal points of commonplace items, and I added to their interest by making the cover plates out of the roots and burls of cherry trees (photo facing page, bottom left). They are wedged in place with slivers of wood or paper—no screws here either.

My methods of detailing were becoming more flexible, and about this time I applied to the National Endowment for the Arts for a grant. They were offering grants of $5,000 for what they called one-of-a-kind, handmade houses, and I was awarded one to help me finish the interior woodworking details. The grant also helped me develop some of the special wood-bending details I used on some lamps (photo facing page, bottom right) and in the kitchen.

Kitchen detailing—I'd done a lot of custom kitchen cabinetry during my contracting days. When it came time to do ours, I had plenty of experience to draw from. We didn't want a lot of space. We laid out the kitchen so that the work surfaces, the cabinets, the pantry and the phone would surround the cook. Almost everything can be reached from the center of the floor.

I had worked with laminated curved panels in some of my previous commissions. The soft rounded forms go well with the functional demands of a kitchen. For example, overhead cabinets that curl under give you more elbow room at the sink and counter. To make them, I resawed figured pieces of cherry into thin bookmatched strips, which I then bent with a hot iron and glued up in layers over a form, a process described in detail on p. 133. The result is a wide, curved and rigid panel with a repeating grain pattern. The curved panels lend a soft, sculptural appearance to the cabinets. The profile is also especially well suited for storing our handmade plates and saucers individually on edge.

The overhead doors slide so that there are never any corners to bump your head on. The problem was figuring out how to hang sliding doors with curved bottoms. I tried many different methods before coming up with a good solution—supporting the door from its bottom edge with a modified aluminum track that fits into a slot at the base of the cabinet (drawing, p. 132). The door handles are an integral part of the frame for each bent panel.

Conventional base cabinets have always irritated me. What you need always seems to be on the bottom shelf near the back. Jeanne and I decided that if the storage space was inconvenient and inaccessible we didn't want it, so I set out to fill all the space under the counters with drawers. Curved fronts work well here, too. We store round lids vertically in slotted compartments. The curved bottoms also make the drawers easy to wipe out.

Below the fixed panel directly in front of the sink, I built a shallow drawer that just clears the plumbing. This is where we store tall botties of detergent and cans of cleanser. What looks like a bottom drawer pulls out, but has no bottom. The dishtowel rack is in here, and this is where we keep sponges and other things that are usually put away damp.

The inside corners of base cabinets can be storage nightmares. Lazy Susans are helpful, but if you have a large one, you still have to shuffle around between the shelves. My solution was to build a lazy Susan on which the shelves revolve independently. This way we don't have to spin the entire lazy Susan and all its contents to get to something.

Our underground tunnel, where we also store food, runs right under the kitchen counter, and I left a hole through the slab for a dumbwaiter. I built an inconspicuous guillotine-type hatch into the wall just to the left of the sink. The bottom of the dumbwaiter pulls up even with the countertop so that we can slide bottles directly onto the conveyor.

The 16-in. space between the refrigerator

Interior details. **A massive fireplace built of local stone, top, is at the core of the house. To build the cantilevered, curving couch, Sollman drove 1x8s 7 in. into the fir columns in the wall, then trimmed their ends to the curved shape of the drawers. Because there was a compound curve at the inside corner, center, the gaps between the bent cherry strips had to be filled with cherry wedges called darts. The laminated bending technique used for the kitchen drawers and sliding doors was also applied to the drawers in the couch, and to lamps, like the one at bottom right. Trim steps around and emphasizes the switch plates and outlet covers, bottom left. Made of cherry roots and burls, these are held in place with small wedges of wood or paper.**

From *Fine Homebuilding* magazine (June 1983) 15:75-79

The kitchen **is fairly small, but carefully planned and very efficient. All of the countertops (top) are unfinished maple chopping blocks, and there are specialized drawers instead of storage cabinets under the counters. The curved cabinet doors slide so that they will never be in the way, whether open or closed. Their shape also works well for storing plates or dishes on edge in individual slots (bottom left). Sollman had to design the detail shown in the drawing below because there are no commercially made sliding tracks for curved doors. The telephone has its own booth (left), with a door that slides down and in to act as a writing shelf. There's a small bar above and a trash bin below. The door at the end of the counter leads to the pantry under the entry stairs, and the panel in the wall covers access to the dumbwaiter. The kitchen knives are stored together (bottom center), each in its own modular block.**

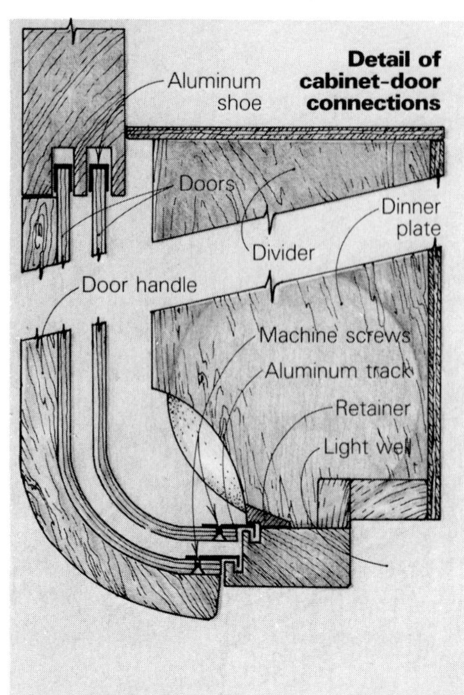

and the pantry door left just enough room for the phone and the garbage can. (When I think about it sometimes, one makes a great container for the other.) I decided to build the phone into its own booth (photo facing page, center), with a door that could be pulled down and pushed into the compartment to provide a small writing area. As the lid opens, a small light comes on to illuminate the phone and writing surface. A small bar pulls out above the booth. Below are two deep drawers for long items like rolling pins. I built in a garbage can. It's simply a box with a raised wooden edge, sized to accept a standard plastic trash bag.

From the outset Jeanne and I decided that carrying a cutting board around would be inconvenient. We would chop and slice stuff right on unfinished maple countertops. I built and installed them, then took my meat cleaver and hacked them all over. Like getting your first scratches on a new car, it's good to do this in a hurry so you don't worry about it any more. Near the middle of the counter I cut a 3-in. by 4-in. hole that a tapered block fits into. Vegetable scraps go down the hole into a plastic pan in the drawer below, from which they're emptied into the compost pile.

Under this counter, I built a knife drawer with removable modular blocks into which each knife fits (photo facing page, bottom center). This keeps the knives handy and protects their edges.

Some of the bottom drawers are large bins for flour and dog food. Their rounded bottoms make it easy to scoop out the contents.

A cantilevered couch—I took the wood-bending procedure one step further when I built the couch in the living area. I had planned to build the couch on the floor, but decided to combine the curved bottom with cantilevered support. I introduced a compound bend that sweeps under the couch and follows the contours of the stone wall behind.

The 1-in. by 8-in. supports for the couch fit into 7-in. deep slots in the fir columns. I cut them a few inches long, applied lots of glue and then drove them in with a sledgehammer. Then I trimmed away the battered ends to match the contour of the drawers below. I cut 3/4-in. birch veneer plywood to shape as a platform for the cushions, and attached a lipping to retain them. I bonded the inside curve to the plywood in pre-bent layers. The first two were glued and screwed. The rest were just glued and clamped.

To bend the drawer fronts, I followed essentially the same procedure as I did to bend the kitchen-cabinet doors, except for the two inside corners. For this compound bend, I had to make a separate form. Of course, when the individual strips of cherry were laid out across the form, they spread apart at the bottom of the curve. I filled these gaps with darts (wedge-shaped strips of wood). Since a drawer would intersect an inside curve, this cutline had to be laid out very carefully. The finished drawers are supported from their sides by the cantilevered 1x8s. □

Bending cabinet doors

After I decide on the size and shape I want my curved panels to be, I build a bending form. It is an inch or two larger than my largest panel, and I make it very carefully. I check it against a template to be sure that there aren't any bumps or dimples in the form, because they would prevent the plies of the panels from bonding. The laminae of the panel are held down with straight wooden ribs that run across the form. The ribs are secured with right-angle threaded rods that fit into reinforced holes in the side of the form.

1. I begin by resawing 8/4 (2-in.) cherry stock into 3/32-in. thick bookmatched veneers and sanding them.

2. I bevel the edges of the top layer of thin veneers on the jointer, so that when they butt, I'll have a V-groove. They are held at the proper angle with a simple jig I made.

3. Next, I lay out the first layer of strips next to the jig, and mark the points at which the curve will begin and end.

4. The middle layer of strips spans the joints of the two outside layers. I mark each successive layer in the same way **(A)**. I take my time, and make sure that the edges fit closely and that I have chosen the first two plies carefully so that the grain patterns of each ply are not parallel. This will decrease the chance of the finished panel splitting with the grain.

5. I pre-bend each strip on a homemade iron, heated with propane **(B)**. I use a piece of pipe that has a slightly smaller radius than my form, and bend between the lines marked on each veneer. The iron gets very hot and often scorches the wood a bit, but this is the back side so it doesn't matter. I don't use any water, because it is apt to warp the panel as it dries. This process makes it easy for me to lay the strips over my jig without soaking or snapping them.

6. Next, I staple a plastic sheet to the form to keep the panel from sticking to it. I lay out the first layer of strips, then apply glue to the second layer and lay them over the first **(C)**, topping off with another sheet of plastic.

7. Starting at the middle and working toward the ends, I clamp the two plies together. I work fast but carefully so that the bottom strips don't spread apart. Using push-pins along the panel's edges helps to keep everything lined up. I've tried to laminate all three layers at once, but trying to keep everything in place drove me crazy. I work the curve last, then tighten all the ribs on the jig. After about two hours, I remove the ribs and the top plastic sheet, scrape off glue squeeze-out and apply the final bookmatched veneer the same way.

8. Two hours later, I remove the ribs and scrape away all the glue that has oozed out between the pieces. This has to be done while the glue is still soft enough to remove easily. After the surface of the panel is clean, I replace the ribs and leave the panel to dry overnight, removing the plastic sheet so the moisture in the glue can evaporate. The next morning, I sand the panel smooth and trim it to size.

9. I've modified a standard sliding glass-door track and shoe to use with the curved sliding doors. I clamp it to the panel, drill countersink holes, drill through with the bit **(D)** and then tap each hole. I then screw the aluminum track to the panel with machine screws. Last, I file the shanks flush to keep the sliding doors from interfering with each other. *—P. S.*

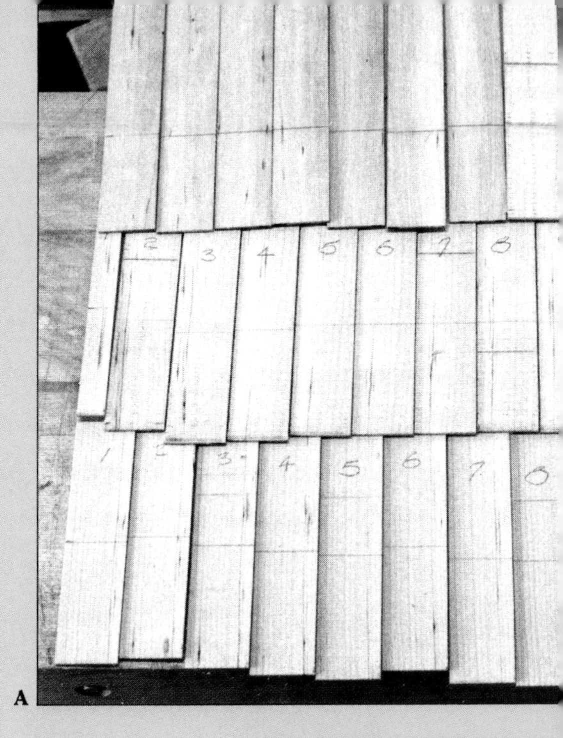

A

B

C

D

A Craftsman Studio

Heavy timbers joined by hand-wrought steel bands

by Charles Miller

Architect Gordon Lagerquist comes from a family of carpenters. Before immigrating to the United States, his grandfather Johannes worked as a carpenter in the heavily forested regions of central Sweden. Drawn to the new world, Johannes Lagerquist settled in Seattle, Washington, and raised a family of sons who also became carpenters.

Like his father, Arthur, and his uncles, Gordon Lagerquist learned carpentry skills at an early age, and before he was out of high school, he knew how to put together a building. But it wasn't the physical work of erecting a structure that captivated Gordon Lagerquist—it was the work of envisioning the spaces and developing a cohesive set of details that appealed to him.

After twenty years in the profession, Lagerquist recognizes the debt that he owes to his carpenter ancestors. But his designs have also been influenced by the shapes of logging-camp shanties, the net lofts and warehouses of the local fishing villages and the restrained demonstrations of carpentry skills common to West Coast Arts and Crafts style bungalows. One of his most recent projects, a studio for a painter on San Juan Island off the coast of Washington, is a good example of how Lagerquist combines imaginative structural details to make his architecture.

Away from the house—Lagerquist's client wanted his office and studio to be a structure that stood apart from his home in Friday Harbor. The studio is a companion piece to the client's main house, which Lagerquist designed eight years ago. In that house, the structure is largely composed of heavy posts and beams connected by custom-made metal straps and plates. The metalwork wasn't anything special, but it imparted a distinctive quality to the interior that Lagerquist thought worthy of further exploration.

A couple of years later, Lagerquist saw a show of wrought-iron light standards at the American Institute of Architects office in Seattle. The pieces were made by Jim Garrett, a Seattle sculptor/blacksmith. Leafing through Garrett's portfolio, Lagerquist realized he'd found someone who could make structural steel components with unmistakable artistry. Garrett's work reminded Lagerquist of the wrought-iron gates and railings that distinguished the Craftsman style work of the Greene brothers. He hung on to Garrett's card, hoping an opportunity would arise where they could collaborate on a building. The chance finally materialized when Lagerquist convinced his client to build the painting studio with hand-wrought metal straps banding together the timber components.

The original studio plan called for one rectangular room, 12 ft. by 18 ft. But it soon grew into a 24-ft. by 28-ft. space that included a bathroom equipped with a shower, and two small storage closets (drawing facing page). One closet is for the storage of canvases; the other opens on to the front porch and houses gardening tools and supplies.

Lagerquist chose a cross-gable form for the studio for two reasons. First, it's a classic arrangement of roofs in the tradition of Craftsman style buildings, and second, the double gables provide an even source of natural light inside the studio. The high windows in the upper gable are kept small on the southwest side and quite large on the northeast side to let in the indirect light so prized by painters.

To stay within a zoning-decreed height limit of 21 ft., the ceiling height at the peak of the upper roof is 18 ft., and it slopes downward at a 4 in 12 pitch to 15 ft. The peak of the lower roof is at 13 ft. and slopes down to 10 ft. at the exterior wall. The spatial organization is open and equilateral, which intensifies the focal point at the center of the square plan (photo above).

Perched on boulders—The studio sits like a raft, floating above a bed of beach cobbles on a foundation reminiscent of Scandinavian and Japanese timber-frame buildings. Lagerquist designed the studio to sit on a framework of 10x12 timbers cantilevered beyond six piers. But these aren't ordinary pier blocks—they're 5-ton granite fieldstones. Contractor Dave Smith says the stones came from a ranch at the other end of the island, where they served only as obstacles. The ranchers had been clearing some forested land for livestock, and with a tractor, they'd rounded up a pile of boulders. They were happy to get rid of the rocks, and in fact were incredulous that somebody would actually take some of them away without charge.

Lagerquist looked at every available boulder before deciding which ones to take. Smith's crew used a hefty front-end loader to lift the stones into the bed of a 10-yd. dump truck. To protect both truck and stones, they put a 1-ft. thick layer of earth in the bed of the truck to absorb any sudden shocks should a stone slip out of the loader's bucket. Meanwhile, back at the site a backhoe was digging six craters for the stone piers.

The crew used a crane and a sling to place the stones. After the first one was in position

From *Fine Homebuilding* magazine (February 1989) 51:36-40

The principal painting space, above, is about 12 ft. square, and bracketed by timber columns. Shojis that slide into a pocket between storage closets (drawing below left) are paneled with vinyl-covered cork, and can be used to control the light or to function as bulletin boards.

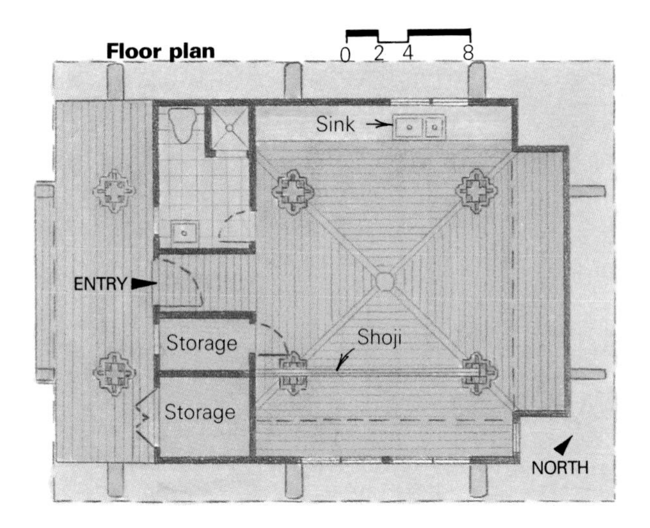

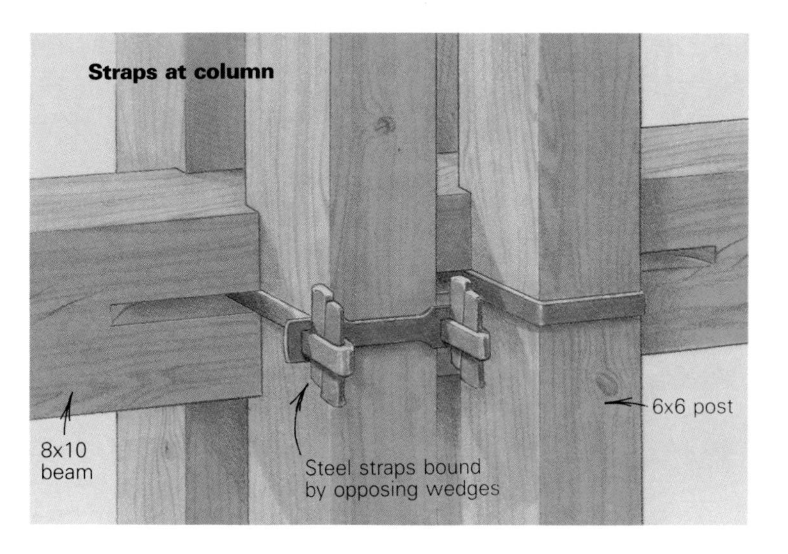

The timber trusses were fabricated in the framed lower portion of the studio, then lifted into place by a crane. Note how the upper gable is perpendicular to the lower one, bearing on beams that pass through the post clusters.

From the east, the studio hovers above a deep shadow and a field of cobbles. The floor framing is mostly 6x12s, placed atop three 10x12 girders. The tenon of one of the 6x12s can be seen protruding to the right of the stairs. Where upper and lower roof truss chords meet, they are secured by steel straps made by Seattle metalsmith Jim Garrett. The straps are tightened by driving opposing wedges into slots formed into the ends of the straps.

at the right height, they used the top of it as a datum point for the transit and set their batter boards and stringlines to that level. Then they lowered the next boulder into its hole. After checking it for level relative to the datum stone, they lifted the boulder so that the hole could be excavated or filled as necessary. Working for eight hours, the crew was able to set the six boulders within ½ in. of level.

To get solid bearing, the crew pumped concrete around the base of each stone, filling any voids created during excavation. Then they drilled a 1½-in. dia. hole in the top of each stone, 8 in. deep, for 1-in. dia. drift pins. Crew foreman Dave Diffner says they used a gas-powered hammer drill to excavate the holes. Each hole took about 20 minutes to drill, but then another 20 minutes was needed for the operator to recover from holding the tool.

Bearing plates made of ¾-in. steel plate sit on the stone piers (drawing facing page). They are embedded in a layer of expanding grout to bring them all to the same level and are mechanically tied to the stones by 1-in. drift pins that are welded to the bottoms of the plates. The pins extend 6 in. into the boulders and are secured with epoxy.

When the epoxy and grout cured, 10x12 girders were bolted to the plates in order to support the floor. Atop the 10x12s and perpendicular to them are five rows of 6x12 beams, which are spaced equally and which divide the floor substructure into four bays, each a little less than 6 ft. wide.

Old-growth timbers—Lagerquist knew that the timbers for walls and roof would have to be free of heartwood to ensure that they wouldn't twist as they dried. Washington's Olympic Peninsula is one of the few places in the world where you can still find free-of-heart 30-ft. long 10x12s. The timbers for this job came from the Windfall Lumber Company in Sequim, where they were cutting first-growth Douglas firs near the Soleduck River. As the timbers were being unloaded, Diffner remembers being dumbstruck by their perfection. "Some of the pieces had 95 growth rings to the inch," he recalls. The wood was so stable that it required no truing to make it square.

Diffner and his carpenter Larry Otto used 6-in. Makita planers to surface the portions of the timbers that would be exposed inside the studio. Where timbers emerge on the outside, they were left rough.

Work began in the early fall, and the contract called for the job to be done by year's end. With the rainy season bearing down on them, the crew put up the portion of the studio that would be stick-framed and fastened some temporary rafters over it to hold aloft a plastic tarp. This gave them a drafty but dry makeshift shop in which they could cut the timber-frame parts.

Diffner made full-scale plywood templates of the top and bottom chords of the timber trusses. He and Larry Otto used a 16$\frac{5}{16}$-in. Makita circular saw and a handsaw to rough out the

Photo top left: Gordon Lagerquist

parts, and then cleaned up the cuts with chisels and planes. A king post made out of a 4x16 connects the ridge to the lower chord and simultaneously joins the top chords by way of hardwood dowels driven through the king-post tenon. At the outboard intersection, the chords are secured by interlocking cuts and Garrett's steel straps (bottom photo, facing page).

Steel straps—The trusses and columns are held together by straps of mild steel. The building official who checked the plans arched an eyebrow when he saw the strapping scheme, but Lagerquist came armed with an engineer's calculations that showed the typical cross section of one of the steel straps to offer three to four times the strength of commonly accepted fasteners and framing plates, so the plans were accepted without delay.

Each strap is composed of two U-shaped bands with eyes at the ends (drawing p. 135). Where they wrap around the edges of the timbers, they are typically let into shallow grooves. Once the eyes in mating pieces are aligned, opposing wedges are driven into the eyes to draw the straps and the timbers into a tightly bound unit.

Over each boulder-pier stands a quartet of 6x6 posts (photo right). The posts fit into square steel sockets that are welded to a steel plate. Holes in the plate correspond to holes in the steel bearing plates atop the boulders. The posts aren't held down by any bolts or nails—the weight of the structure and the combined connections of the framed walls and the beams they carry in common are enough to resist any uplift forces. The steel sockets resist loads from the side.

Using 1½-in. by 3-in. bar stock, Garrett made massive L-shaped feet that tuck into the vertical slots between the 6x6 posts. They are grooved to accept the steel straps that wrap the posts into a unified bundle, and they splay out at the base to end up over the pier-block bearing plates. The entire assembly is secured with ¾-in. bolts that run through the L-shaped feet to the bearing plates. To give the bolts a handmade look, Garrett welded rivet-like tops to the exposed bolt heads. A cross-shaped cover plate fits into the space between the posts. All told, there are 7,000 lb. of custom steel connectors in the studio.

The clustered posts are central to the stacked layers of structure that characterize this building. At the 7½-ft. level, a pair of 8x10 beams running the long dimension of the studio slips through the spaces between the posts, where they bear on inch-deep shoulders cut in the posts. The beams cantilever beyond the footprint of the studio to carry the gable ends of the lower roof. Above them, other beams interlock with the posts to carry rafters and the ridge beam for the lower roof. Two of the trusses that carry the high roof nestle into openings at the tops of the four post clusters.

After three intense weeks of working heavy timbers in a tiny shop, Diffner and his crew

Drawing: Bob LaPoint

Fieldstone foundation

were ready to hoist the completed trusses onto the roof. Typical late fall weather in the San Juan Islands includes a high-pressure ridge that fends off the rains for a week or so. When the expected break arrived right on time, they brought in the crane for truss lifting (top photo, p. 136).

Muted finishes, related shapes—The amber tone of Douglas fir is the predominant color in the finished studio, but neutral grey accents complement the color of the steelwork. Designer Misty Todd-Slack devised the grey, laminate-covered counter that sits at the base of the west wall, concealing flat files, a hot-water heater, a stereo and a roll-out cart that contains the artist's paints and brushes.

In the salty air of Friday Harbor, Garrett's mild-steel pieces would quickly rust. To keep the interior steelwork from oxidizing, Garrett coated it with a clear vinyl called Stays-On (Steelcote, 3418 Gratiot St., St. Louis, Mo. 63103). This material is usually used as a primer instead of a finish coat, but Garrett finds that it holds up well enough without a final color coat, allowing the steel to retain its blue-black forged color. Outside, the steel parts were left to rust.

In keeping with most of the island's houses that were built during the Craftsman era, the studio is covered with cedar shingles. To break up the relentlessly regular pattern of most sidewall shingle jobs, Lagerquist had the crew work chevron-shaped shingles into every other row (bottom photo, p. 136). The entire exterior of the building was sprayed with linseed oil, using a pressure sprayer normally used for cleaning truck engines. The linseed oil will need to be reapplied every five years to retard discoloration and deterioration of the fir and cedar.

The keystone shape of the king posts became a repetitive element in the final detailing of the studio. Dave Marsidon made clerestory windows with muntins that flare out at the same angle as the king posts, and Lagerquist teamed up with metalsmith Garrett to design inverted-pyramid light fixtures that echo the king-post line (photo left).

Looking east, the full 18-ft. height of the studio space can be appreciated. The keystone shape of the king posts is repeated in the window muntin pattern and the light fixtures.

Epilogue—Lagerquist's studio for an artist sits like a wooden geode, its exterior darkening with age and the metal bands gaining a crusty coating of deep orange. Inside, the light seems to come from everywhere, with faceted highlights and shadows emphasizing the beamwork. But one aspect of the building didn't age quite as well.

About a year after construction, the junction of the drywall at outside corners started to shear. Alarmed, the crew came back to put a transit on the wall to sight for any deflection. Sure enough, the bold 6-ft. cantilevers carrying the long walls had sagged—up to an inch in some spots. Lagerquist surmises that three factors combined to cause the deflection. One, the 10x12 beams that carry the cantilevered section were still green when they took the load of the framed wall, and green timbers under these conditions can sag, or "creep" as it's known. It doesn't threaten the safety of the structure, but it affects the look of it.

Second, as the timber frame lost moisture and the members shrank, some of the load from the high roof seems to have been transferred to the clerestory framing and from there to the framed walls atop the cantilevered beams. This extra load only made the situation worse.

Third, as a result of some miscommunicated instructions, 2-in. deep notches were cut into the 10x12 beams over the piers. It was the 6x12s that were supposed to get notched. This reduced the carrying capacity of the cantilevered beams to that of 10x10s, and caused them to sag under load.

The fix turned out to be as elegant as the initial footings. Using jacks under the protruding 10x12s, the crew slowly lifted the sagging walls back to level. Then they placed another row of glacial boulders and bearing plates under the outside walls. □

Charles Miller is senior editor of Fine Homebuilding. *Photos by Gary Vannest except where noted.*

A General Remodel and a Major Addition

A 1920's bungalow gets stripped down, spruced up and doubled in size

by Scott W. Sterl

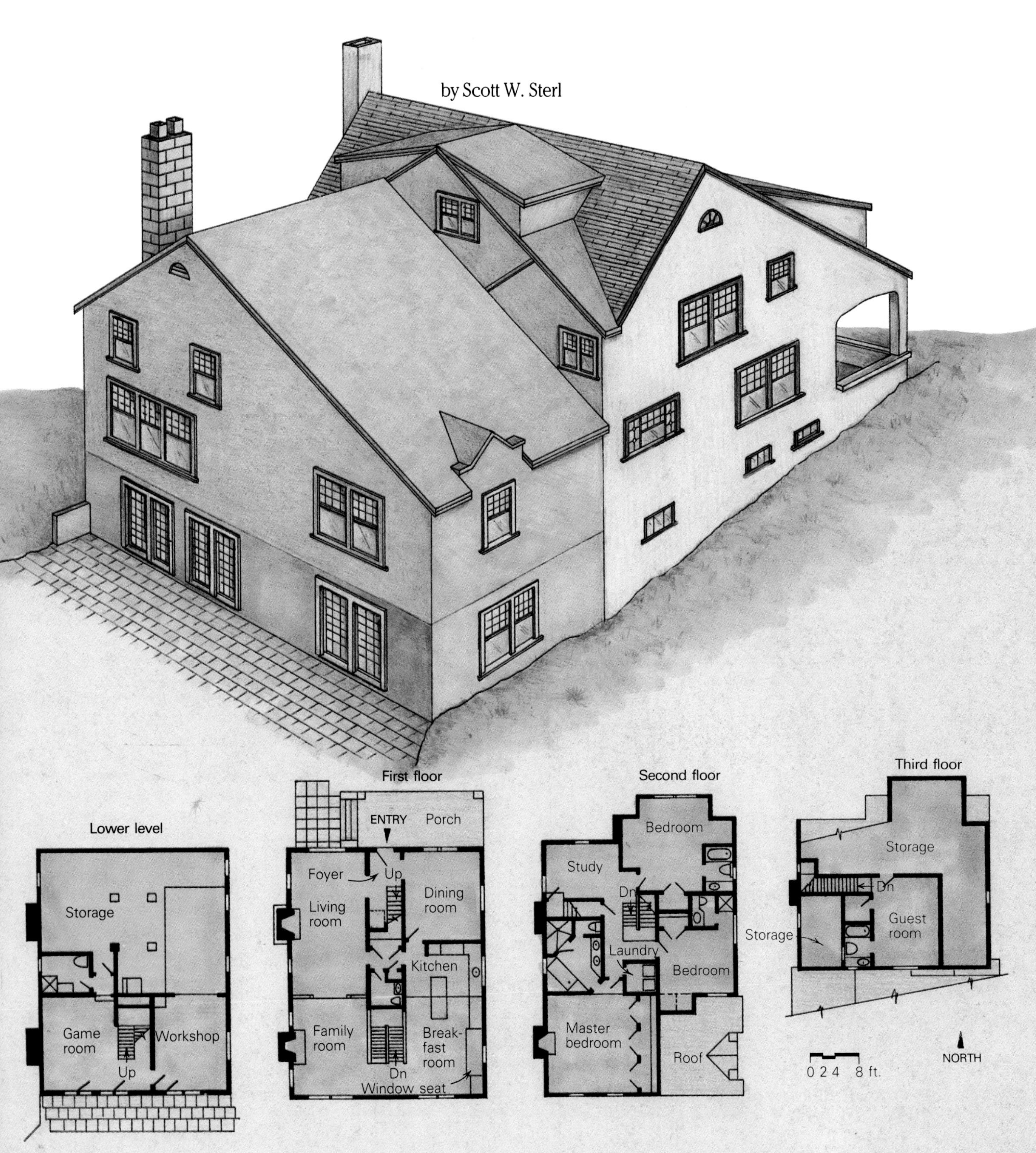

It was a typical August day in Washington, D. C.: 100° F and 100% humidity. Informally attired guests wandered through the home of Adam and Laurie Sieminski, which was soon to be the object of a major remodel and addition. The Sieminskis had invited a number of their friends to witness the "before" condition of this three-bedroom, 2½-bath 1920's bungalow-style house.

I was included in the group, along with Tom Gilday, president of the design/build firm I work for. Gilday had first met the Sieminskis several months earlier. They had been looking at houses for two years and had finally found something with potential. The couple placed a contract on the house with the stipulation that they be given ten days to contact builders and architects to discuss remodeling options. They met with three firms. One said that they were so busy they couldn't begin the design work for six months. Another said that what the Sieminskis wanted—a large addition tastefully integrated into the existing house—couldn't be done. Our design/build firm got the job.

Humble beginnings—The Sieminski house is situated in an architecturally eclectic neighborhood, with homes ranging in style from Georgian to 1960's contemporary. A deep, but undistinguished front porch, a shed dormer with false half-timbering and a parapeted gable over a second-floor window were the only distinctive features facing the street. Unfortunately the stuccoed exterior and all the trim, including the false half-timbers, were painted the same dull beige.

Inside the house, the foyer was flanked by the dining room and living room, with the kitchen in the rear. A sunroom off the living room led to an amateurish deck. The interior trim was painted and plain. A keystone arch in the foyer was the only special detail, but all it framed was a closet and a dark narrow stairway leading straight to the second floor. Upstairs were three bedrooms, two of which shared a bath. The master bedroom, which was above the living room, extended the depth of the house. An unfinished attic was accessed by a pull-down stairway in the hall.

The Sieminskis wanted to remodel the original house completely and to construct a large (18 ft. by 35 ft.) 2½-story addition. The addition that the Sieminskis foresaw would almost double the size of the original house. Because the lot was long and narrow (300 ft. deep by 50 ft. wide), the only location available for the addition was at the rear of the structure, which meant eliminating the sunroom and deck.

The plan was to expand and update the kitchen and add a breakfast room, a family room, a workshop and a game room. Laurie is a quilter who also teaches, so the game room would double as a quilting classroom.

Second-floor improvements would include separate baths for the children, a family study, a laundry room and a master bedroom with bath. The attic would be expanded to contain a guest room with private bath (see floor plan, page 139).

Assembling the masses—With an addition of this magnitude, I preferred to blend rather than contrast with the exterior style, especially as the extension was to be the same width as the existing house. This presented some obvious design elements for me to start with: stucco as the finish material, 12-over-1 double-hung windows, exposed rafter tails and gable rooflines. But the task of how to blend the massing of the new and old seemed monumental at first. After some study, however, the solution came to me rather quickly.

The ridge line of the existing house was parallel to the street (drawing, page 139). The obvious way to add on at the back of the house was simply to extend the roofline like a saltbox. But I quickly dismissed that possibility because it wouldn't encompass enough additional room, and it would have looked boring. Instead I added a visually rich collection of rooflines.

I ran the main ridge of the new addition perpendicular to, and 4 ft. below, the original. This roof shelters the new master bedroom. Because this room is only half the width of the house and aligns with the west wall of the family room, the roof peaks at the center of the bedroom and then extends to the plate line at the first floor breakfast room. I broke up the long slope of this roof plane with a parapeted gable over the kitchen window seat, matching the one on the front the house.

I created space for a third-floor guest room by running the ridge of another gable roof—this one with a 3-in-12 slope—at the same height as the original ridge. Rake boards were applied to the new exterior wall at an 8-in-12 slope so that, from the outside, this bedroom appears to be an 8-in-12 gable roof with a pair shed dormers on either side.

Quoining, a phase—The Sieminskis wanted a fireplace in the family room and in the master bedroom above it. The chimney for the existing fireplace is built of brick, inset with three stucco diamonds at seemingly random spacing. In order to minimize the visual impact of the new two-story fireplace, I decided to reverse the materials. That is, to use stucco, which would dissolve into the house, for the main structure and employ brick insets. As I was reversing the materials, I elected to reverse the spacing of the brick diamonds as well (top photo, page 141).

Only superficial changes were made to the front of the house, but they dramatically altered the streetside view. The greatest change was effected with some blue-green trim paint.

To enhance the exterior further, I introduced brick quoins as a new detail. Quoins (from *coin*, the French word for corner) are generally a decorative interlocking of bricks or stones at the corner of a building. They first appeared in Renaissance architecture, but over the years have been used in other styles as well, among them Georgian and Beaux Arts. I felt the quoins were necessary to define the corners of both the original house and the addition, as well as to contain and define what was getting to be a pretty large mass.

After mocking up various designs by laying out bricks on the plywood subfloor, we decided to install the quoins in an alternating pattern of 8-in. and 16-in. legs, each one three courses high, to lend rhythm and a less formal feeling than quoins of the same size. A column of 8-in. wide quoins was also placed at the juncture of the new and old parts of the house, subtly delineating this intersection.

We actually used brick soaps—½-in. thick bricks—to make the quoins because we were applying them over the existing stucco, as well as setting them in the new. On the corners of the addition, 2-ft. wide sheets of cement backer board were installed on top of the plywood. The soaps were then applied with PL-400 construction adhesive. We mortared the joints and painted them to match the stucco. On the existing house, the soaps were simply glued on top of the stucco with PL-400.

Paint makes a difference—The covered front porch had little character. The brick columns at its two outside corners, which supported the stuccoed Tudor arches, had been painted the same color as the stucco, destroying whatever architectural interest they had added previously. My plans to improve the porch were simple. I specified a 2x6 band to wrap the columns at their intersection with the stucco (photo left), hiding this awkward joint.

All trim, including the exposed rafter tails and the false half-timbers, was painted a vibrant blue-green. This was an easy step and yet made perhaps the greatest difference in the appearance of the house from the street.

Three kinds of wood—The Sieminskis requested at least three kinds of wood in the interior trim. We presented them with an integrated interior trim package incorporating maple, cherry and walnut as the major woods, with a sprinkling of mahogany.

The interior trim falls into three categories: baseboard and casing; stair railings; and fireplace treatments. New Era Cabinets in Frederick, Maryland, fabricated all the custom trim, which was installed by Frank Johnson, one of our best carpenters. The baseboard on the first floor, in the second-floor public areas and

Drawings: Michael Mandarano

From *Fine Homebuilding* magazine (December 1990) 64:38-42

in the lower-level game room is 4¼-in. wide maple, with walnut shoe molding below it and a cherry bullnose on top, set off by a ¼-in. groove routed in the top edge of the maple. The cherry was stained with mahogany stain to enhance its reddish color, and the maple and walnut were finished with clear polyurethane.

At the openings that flank the foyer—one to the living room, the other to the dining room—the base moldings are mitered to become the door casing (bottom photo, page 142). This accentuates the openings and makes this space more important. Where the side casings meet the head casing, squares of walnut were inlaid along the miter joint.

The rest of the interior window and door trim is a stock 3½-in. casing with eased edges, customized by routing a groove near the outside edge. It relates to the custom trim and was an inexpensive way to make it unique.

Diamond-studded newel—The most dramatic change in the layout of the existing house was the reconfiguration of the main stairway. Originally, the impression from the front door was of a confined passage to the second floor. We removed the existing straight stairs altogether and installed a switchback, (or dog-leg) stair, with a landing two-thirds of the way up and a railing around the second-floor opening. This had two effects. First, it opened up the second-floor hall area, which had been dark and narrow. And second, the change in direction of the stairs allowed entry to the second floor at the study, which is a more public area, rather than in front of a bedroom door.

I designed a 6x6 maple newel post with a pyramid top and centered it on the new front door (bottom right photo, p. 42). It's set on the diagonal, with the first riser wrapping around it as a base. Alternating vertically up each face of the post are inlaid squares of walnut and cherry. Fabricated by New Era, the 6x6 was glued up from 5/4 stock, and the 5⁄16-in. thick squares were inlaid with a router. John Gage, owner of New Era, cut the pyramidal top on a table saw. The horizontal grooves, which set off the handrail and pyramid top and add interest to the newel post, were also cut with a router.

Around the stairwell on the second floor, 4x4 pyramid-topped newels continue the theme begun on the first floor. Although painted instead of clear-finished, they sport a pattern of vertical and horizontal grooves. The handrail is 3¼-in. by 1½-in. maple, doweled into the newels. The square balusters are set on the diagonal between two horizontal cherry rails (also doweled into the newels). The newels and balusters were primed in New Era's shop with a lacquer-base primer, then installed and finished on the job. The pre-priming simplified the joining of stained and painted parts and provided some protection during the finishing stage of the construction.

The stairway to the lower level, which contains the game room and workshop, was placed between the breakfast and family rooms. Open to both rooms, it's protected by a balustrade like the one on the second floor (photo right).

New and old chimneys. **To minimize the visual impact of the new chimney, yet still relate it to the original, the architect specified a stucco finish with brick diamonds, which is just the opposite of the original. Half-inch thick bricks, called soaps, were applied to the corners [illegible] building to give the effect of quoins.**

Expanded kitchen. **The kitchen is organized around a king-sized island with a Corian countertop and cherry nosing. Although the breakfast area isn't physically separated from the kitchen, its space is defined by a cathedral ceiling over the table. The railing visible in the foreground, which is just like the one on the second floor, guards the basement stairs.**

Dividing spaces. **Half-walls and maple columns mark the line between the living room and family room. The paneled effect on both the half-wall and on the fireplace mantel was created by simply applying 1x stock over drywall, with the cherry accents adding a touch of elegance.**

Defining spaces—The living room and family room are at opposite ends of what is essentially one long room (photo above). And because the Sieminskis wanted a view from the living room out the family-room windows to the backyard, I couldn't separate these spaces with a wall. Instead, I used a pair of paneled half-walls with maple columns on top of them. The walls feature 1x pine over drywall to create the paneled effect. The cherry bullnose employed on the baseboard here becomes the "capital" on the columns.

I also wanted the living-room and family-room fireplace treatments to be distinct from each other. The one in the living room involved disguising the existing brick surround and mantel. I decided a vertical treatment would break up the 26-ft. length of the room and also attract attention to the fireplace, which is immediately visible from the foyer. I prescribed a floor-to-ceiling drywall structure, totally enveloping the existing surround. Over this, Johnson nailed and glued vertical mahogany lattice strips, with intersecting horizontal pieces near the ceiling and just above the fireplace opening. This arrangement plays on an Arts and Crafts pattern.

The arrangement of the strips left a framed rectangle of drywall that cried out to be filled. We designed two stained-glass panels—one for Adam and one for Laurie—which Gilday presented as gifts. The panels were mounted on short lengths of copper tubing that hold them out from the wall surface about an inch. Low-voltage lights, recessed in the ceiling

Baseboard becomes casing. **At the openings that flank the foyer, the three-piece baseboard is mitered to become the door casing. Diamond-shaped walnut inlays decorate the upper corners of the casing, as well as the newel post.**

Pyramids and diamonds. **Tucked between the pantry and the planning desk, a window seat (above) provides a place to escape; the columns flanking either side sport pyramid tops. This geometric theme starts in the foyer where the casing features diamond-shaped walnut inlays in the miters (below left) and the newel post wears both walnut inlays and a pyramid top.**

above them, project the colored patterns on the painted surface of the wall behind.

The surround of the family-room fireplace matches the half-walls, with pine lx3s attached to the drywall to resemble frame-and-panel construction. It is capped by a 2-in. thick bullnosed cherry mantel.

Nosing around the kitchen—Judy McCandless, an interior designer who also works for Gilday, designed the kitchen (as well as the bathrooms). Laurie knew that she wanted a large island at which she could bake with her children, hold cooking demonstrations and cut fabric for her quilts (bottom photo, p. 141). Her stipulations on finishes included Mexican tile floors (which would carry through to the family room), Corian countertops and simple white wood cabinets.

Judy called for the oak kitchen cabinets to be finished in white stain so the grain would show through. They were fabricated by New Era, who also fabricated the cherry bullnose for the Corian countertops. The same bullnose borders the valance, concealing cove lighting in the kitchen and breakfast room.

Judy and Laurie were considering ceramic tile for a backsplash, but I suggested using Corian there as well and scoring it in a 6x6 grid pattern, which is what we did. This way we avoided introducing a new material and instead exploited the properties of the Corian, using it in an unexpected way. New Era employed a computerized panel saw to cut the grooves. Judy and I had the electrician adjust the location of his receptacle boxes so that they would be centered in the 6x6 squares.

Judy and I defined the area of the breakfast room, which is not separated physically from the kitchen, by framing a cathedral ceiling over the table. We broke up the long kitchen-cabinet wall by placing a window seat, at Laurie's request, between the pantry cabinet and the planning desk (photo above). This is further set off by maple columns on either side, each having a pyramid top and cherry bullnose.

Departing gifts—Fortunately for the Sieminskis, the couple who had purchased their former home were still living in England. The Sieminskis were able to remain and rent from the new owners while the work was being done on their new home.

Construction began in August of 1988, and the long-anticipated move-in day came exactly a year later. At Christmas, the Sieminskis had another party and invited all of the friends who had viewed the house "before." They also included many of the people who had worked on the house, and even gave personalized gifts to some of us. Adam and Laurie presented Frank Johnson with a complete set of back issues to *Fine Homebuilding*. I got a wooden tie made with three kinds of wood—the same three employed in the trim. □

Scott W. Sterl is an architect with Gilday Design and Remodeling in Silver Springs, Md. Photos by Chris Walle except where noted.

Craftsman-Style Beach House

Getting the most from a 25-ft. wide lot and a tight budget

by Tim Andersen

The house that formerly occupied this 25-ft. by 100-ft. patch of sand in Oxnard, Calif., burned down in 1982. It had been a rental house, and its owners, Rolando and Linda Klein, asked me to design a replacement. Their insurance settlement wasn't enough to cover current building costs, yet the mortgage company threatened to call the loan due if they didn't rebuild. To resolve their dilemma, the Kleins decided to build a small house as inexpensively as possible, and sell it when it was done.

The Kleins and I agreed that the new design should have three bedrooms, two baths, a family room and a breakfast room—all somehow squeezed into 1,500 sq. ft. and costing under $60 per sq. ft. (about $15 below the going rate in Southern California). The challenge was to design a comfortable, efficient house that would feel much more generous than its small square footage would normally allow, and one that could be built using conventional materials and methods to keep costs in line.

The zoning straight jacket—Zoning ordinances required our new house to maintain a 20-ft. setback from the street, a 6-ft. setback from the beach, and a 3-ft. setback on both sides. Further, we had to comply with a 25-ft. height limit. Since one of these tiny parcels sells for about $200,000, everything built in this area tends to fill the allowable zoning envelope with as much house as possible. This leaves 6-ft. wide wind tunnels between each box, bisected by a fence down the property line. Zoning also required four on-site parking spaces, two of which had to be enclosed. Parking alone ate up 30% of our lot coverage to begin with. To make usable outdoor space possible in this situation, zoning should allow for a staggered placement of houses on adjacent lots and zero-lot-line development using fire-rated party walls.

In striking contrast to all this mandated confinement was the expansive, wind-racked beach, hundreds of yards deep with waves crashing onto the shore and receding into the blue Pacific. Clearly, the house wanted as much exposure to this wonderful view as possible.

Fitting into the neighborhood—There were many older beach houses in the neighborhood, built as summer retreats in the early part of this

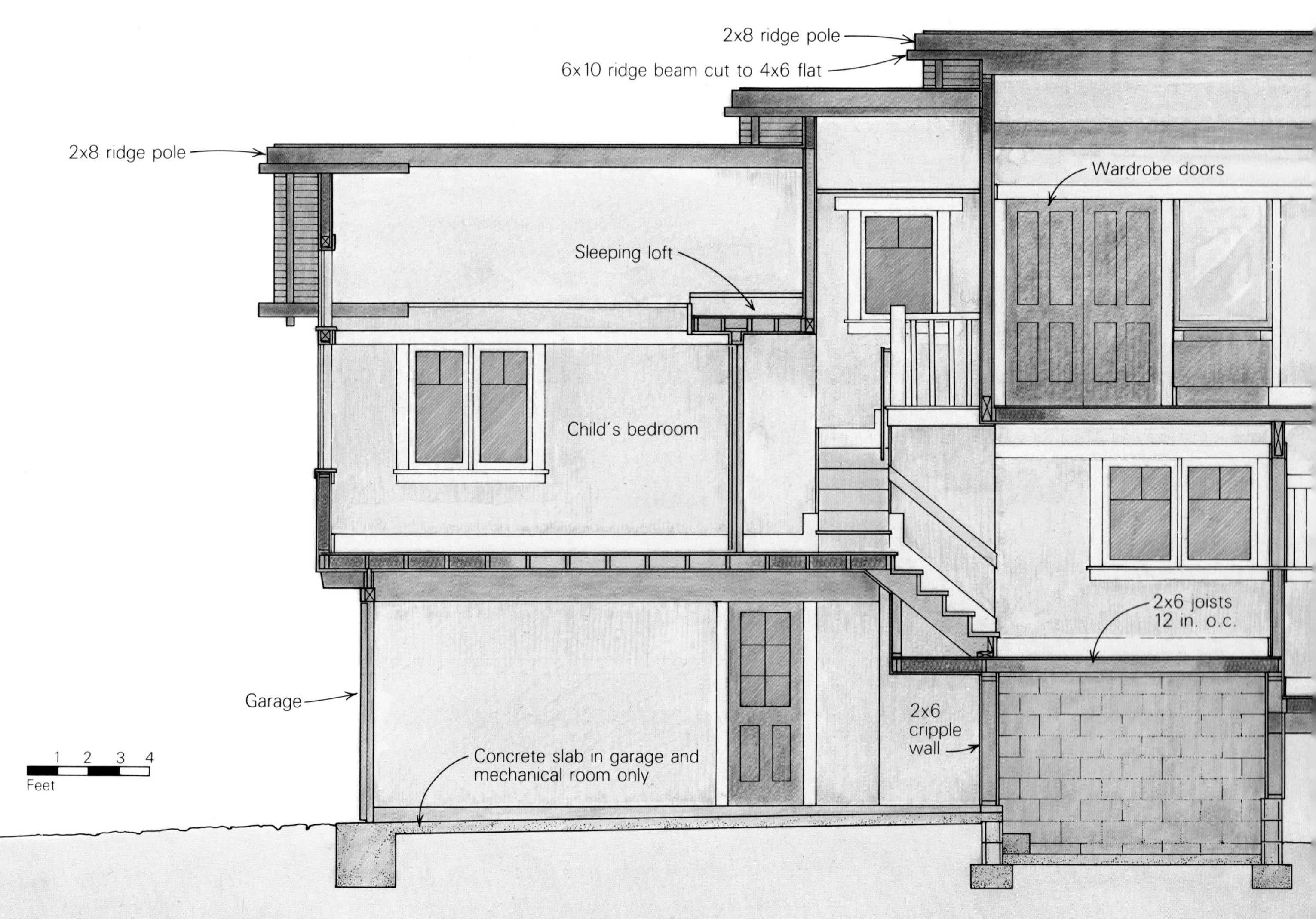

century, when land and construction were cheap. Designed simply for comfort and without pretension, these houses appear to many of us now as archetypes—pure examples of what a beach house should feel like. The older homes are far more satisfying than the new spec houses that are being built with a hook for every potential buyer, each house trying to upstage the others. In my design for the Kleins' house, I wanted to make a clear connection with these earlier beach houses and to avoid the hawking aggressiveness of this newer bunch.

Custom work at tract-house prices—As we sized up the situation, we saw an opportunity to beat the developers at their own game. We would employ the same low-budget construction techniques—wood frame, concrete block, drywall and stucco—but aim for integrity rather than glitz in the final product. Certainly, we thought, there must be buyers who shared our bias. The quality would come in part from careful detailing in design, but the crucial difference would be made by the builders.

Through my work in Pasadena, restoring houses from the Arts and Crafts Period (1900-1915), I had met several contractors and craftsmen who, like myself, found sympathetic precedent in the attitudes represented in this earlier period. Their involvement in a project from start to finish was part of a commitment to this work that placed the emphasis on taking satisfaction in the process and pride in the results. One craftsman I met, who shared my enthusiasm for finding ways to revive this Craftsman attitude and give it fresh expression, was Glen Stewart. We worked together for two years at one of my most fully realized restoration designs, Greene and Greene's Bolton House (see pp. 87-93). With Stewart as the builder, this beach house would be our first chance to bring these ideas into the present.

Coming up with a plan—In response to the narrow slice of beach frontage and the probable construction of another house to the south, I designed the house to be a grandstand to the ocean view, with each space stepping up to look over the one in front (section drawing, below). In gradual increments, the house would climb up over the garage where the two children's bedrooms would be located. These were the only rooms that would not face the beach. Above the kitchen and dining room was the master suite looking over the living-room roof to the ocean. The organization of the interior is clearly visible from the outside as the roofs cascade over one another. The plan has a jog and a notch along the south wall; these not only gain more deck space but they work as well to get beach views from the living-room sitting area and from the breakfast nook farther back.

Because of the tight square footage, it was important to have many rooms share volume and daylighting so they would feel generous. Pitched ceilings and high windows would also help to open up the spaces. The communal areas of the house were designed to be unobstructed by par-

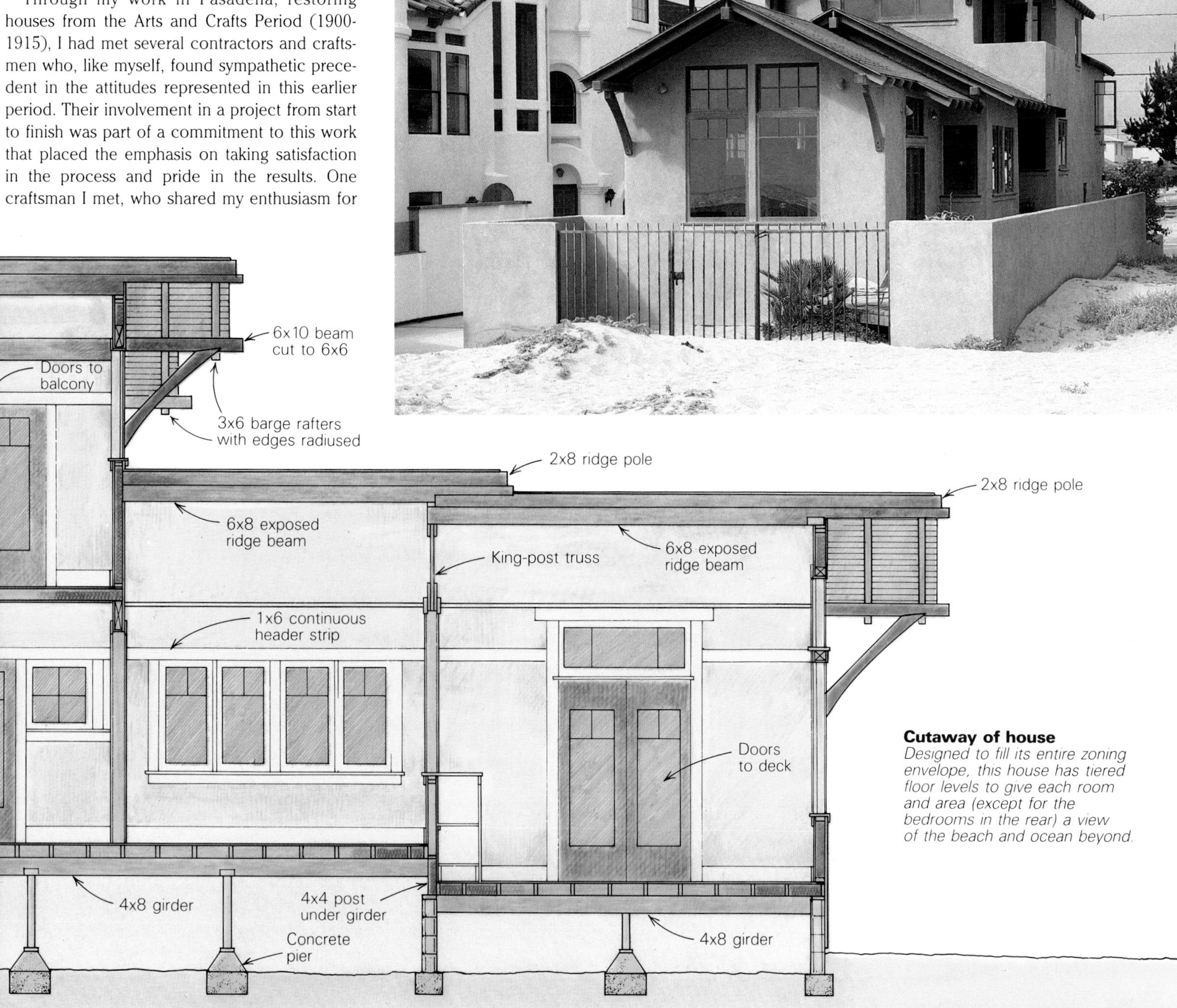

Cutaway of house
Designed to fill its entire zoning envelope, this house has tiered floor levels to give each room and area (except for the bedrooms in the rear) a view of the beach and ocean beyond.

The fireplace (above) sits at 45° and holds the corner between the living room and dining room. Along with the offsets in the plan, half-walls and changes in floor levels, the fireplace is part of Andersen's scheme to create a sense of discrete spaces while leaving the communal part of the house visually open from one end to the other. To make the most of precious space, Andersen used part of the setback for a deck (below left). The jog in the south wall extends the width of the living room and offers a window's worth of view toward the beach. The stairway to the master bedroom (below right) takes a couple of bends and makes this area seem as withdrawn as possible in a small house. The railing at the landing is simple but effective. The rails top and bottom were grooved. The grooves were fitted with spacer blocks glued in at regular intervals, and the resulting holes made mortises for the balusters.

Upper-level plan

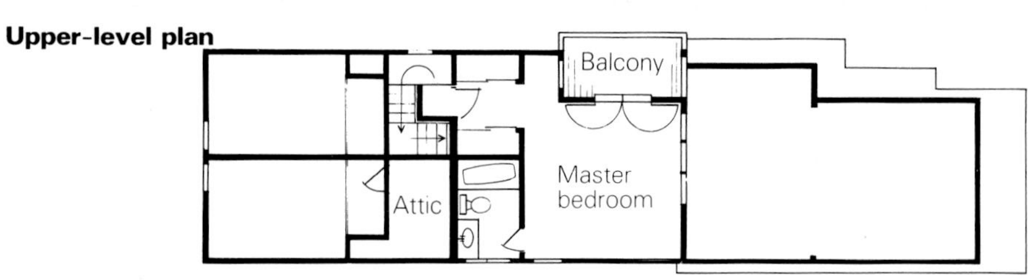

Lower-level plan

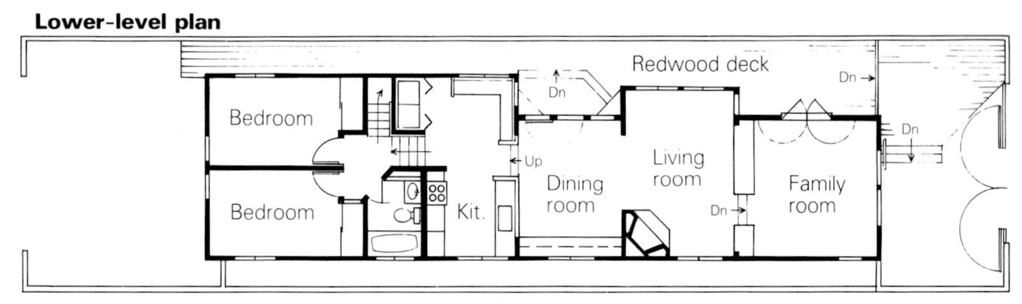

tition walls. You can see from the family room all the way to the rear of the kitchen (photo facing page). Even better, you can look out from the kitchen through the living room and family room all the way to the ocean.

A common problem with many open plans is that spaces aren't clearly defined, which makes it difficult to decide where the furniture should go. Spaces must suggest traffic patterns and a sense of enclosure, yet they need not be walled in to accomplish this. Discrete spaces can be suggested simply by defining the corners, in this case with truss and posts, half-walls and built-in cabinets. As I worked with the plan (drawings, bottom left), it occurred to me that by changing the floor elevations and by offsetting from each other the two sets of steps between the family room and the kitchen, I could create eddies in traffic flow, the obvious areas for seating.

Another concern was to take full advantage of any space not built on for decks and gardens. I wanted the beach sand to come right up to the windows on the west. The northern 3-ft. setback was to be planted in eucalyptus, whose branches would eventually reach the high windows. On the south would be a sunning deck 30 in. above grade (middle photo, left), to be supported by a 6-ft. high masonry-block wall running right along the property line. Anything built above 2½ ft. within the setback would be considered an encroachment. The remaining 3½ ft. of fence above the deck worked as a railing. In this way we could reclaim the precious 3 ft. of wasted space in the setback.

From this deck one enters the house directly into the dining room (photo at top), which I envisioned to be the center of activity. Here, the family and guests could do everything from sorting seashells to savoring coffee. From the dining room the kitchen is two steps up; a half-wall keeps its counters out of sight but leaves the area open to adjacent spaces and to views of the beach. There is a breakfast nook on the same level and a laundry area tucked behind bifold doors. The washer and dryer occupy space over the car hood in the garage below.

The two children's bedrooms have skylit sleeping lofts above their closets. The master suite is up a short flight of stairs (middle photo, right) and around a couple of turns that make it seem as remote as possible in a 1,500-sq. ft. house. The master bedroom has a small balcony, a built-in desk, its own bath and double closets.

The fireplace in the living room has several other jobs besides burning wood. Its placement suggests a corner of the dining room, providing just enough enclosure to define the space. At a 45° angle (photo at top), it faces and anchors the corner seating in the living room.

Foundations—When the permits were secured, Stewart moved from Encinitas, near San Diego, to a temporary place in Oxnard. He assembled his crew and made plans to begin construction in April 1983.

The sand was graded and compacted for conventional, poured-in-place perimeter concrete footings—no caissons were needed. The engineers assured us that the sand had adequate bearing capacity because it percolates so well;

Drawings: Elizabeth Eaton

Looking east from the family room toward the kitchen, the effect of the tiered plan is evident. Two steps up from the family room is the living room/dining room, and two more steps up is the kitchen and breakfast nook. Beyond that, a stairway leads to the bedrooms and baths. By placing each area on a level higher than the one to the west of it, Andersen created a kind of grandstand so that each space has a clear view of the ocean.

Framing, plain and fancy. **The floors and walls for this California beach house are conventional 2x framing (above), except for purlins that extend the top plates of the long exterior walls. These 6x6s are supported with Craftsman-style knee braces. The concrete-block foundation walls were laid up flush with the wood framing and later plastered as a continuous surface. The king-post truss that separates the family room from the living room (below), combines wood joinery and metal fasteners in a design that's both handcrafted and practical. The tops of the two posts are tenoned to form seats for the bottom chords, and the king post and web members are also tenoned on both ends where they are sandwiched between the double chords.**

they also said that if our footings and foundation walls were all properly tied together we would not have any settlement problems. The only slab on grade was to be for the garage and mechanical room. The lowest framed floor is 30 in. above grade, which means that any water flooding the site could pass by the building without reaching floor level. The foundation and perimeter site walls were laid up quickly using 6x8x16 concrete blocks with flush grout joints. These walls would be plastered later, at the same time the framed walls would get their plaster.

Framing—Stewart's three-man crew began framing in mid-April. The platform framing used for the floors and walls was entirely conventional, therefore fast and cheap (photo top left). The more expensive part of the framing was doing the roof with beams and purlins for the open, pitched ceilings. We extended the eaves to their legal limits (9 in. over the setback line) to get the kind of generous overhang that is a hallmark of Craftsman-style houses. The 6-in. rafter tails were radiused with a router using a rounding-over bit, and prestained to contrast with the roof decking. For the roof decking we used ⅝-in. plywood, except where it was visible under eaves. There we switched to 1x6 V-groove T&G decking that is ⅝ in. net to match the plywood. This was also prestained.

The most interesting part of the roof framing was the king-post truss (photo bottom left), which was designed and built by Rodger Whipple. This truss, along with the exposed posts it rests on, visually separates the living room from the family room. The natural wood finish of the truss, posts and ridge beams lends a feeling of warmth and hand-wrought quality to an interior that's mostly drywall.

To support the double bottom chord of the truss, Whipple notched the top of each 8x8 post to form a vertical tenon. One of the 3x8 chord members sits on either side of the tenon, and a bolt through both chords and the tenon holds things fast. The double 3x6 top chords were level-cut to rest atop the bottom chord, and Whipple secured the connection by gluing in a spacer block here; it fits into the gap left between the top edges of the chord and the top of the tenon on the posts. The webs are also tenoned on both ends, and fit in the slot between the double bottom chord and the double top chords. The king post is treated in like manner, though the tenon on top is square in section where it fits into the ridge beam. All exposed edges of the truss members were relieved with a ½-in. rounding-over bit, and the wood was sanded before the Watco oil finish was applied.

Windows and doors—We insulated the interior and exterior walls with R-11 fiberglass batts, and floors and ceilings with R-19. On the interior, these insulated walls help deaden sound.

Double-glazed windows are the common choice in most well-insulated houses. But we wanted a consistent glazing motif with real muntins throughout the house, so we decided to go with single-glazed. The decision to use single glazing meant—according to code—that the allowable glass area for windows, skylights and

French doors could not exceed 20% of the floor area of the air-conditioned spaces. I used every square inch of the allotted area.

The windows were prehung wood casements made from standard parts. They were actually quite a savings over what custom sash and frames hung on site would have cost. The company that built them charged by the number of lites, and their configuration didn't affect cost. The consistent use of the same glazing pattern in windows is a common Craftsman period device that gives a sense of unity to a variety of wall openings.

Stuccoed walls (and a near disaster)—For a sense of unity, I wanted the exterior walls to be consistent in material and color throughout. The finish had to be durable and not require much maintenance. Since cost was also important, stucco was the obvious choice. To reduce labor, we used K-Lath (K-Lath, Div. of Tree Island Steel, Inc., Box 2120, Monrovia, Calif. 91016) so that the building would be wrapped only once. This product includes a vapor barrier, mesh and perforated kraft paper fastened together in one roll.

The concrete-block foundation walls and exterior plywood framing were flush, and intended to blend together without a visible seam. I realized, however, that because of their different rates of expansion and contraction, a crack could develop at the juncture of wood and masonry. The remedy was simply to add a 24-in. wide band of expanded-metal lath that covered the joint and provided extra reinforcement. The block walls took plaster directly in two applications. The frame walls had the standard hand-applied scratch, brown and finish coats.

But once the house was wrapped and given its scratch coat, it looked like a stucco box with funny little windows. It was evident that my aspiring Craftsman Revival house had taken a nose dive, and looked more like a dingbat condo than I dared admit. At this point I decided to reassert our differences with the condo brethren by ignoring the standard range of nondescript, tasteful color coats offered, and instead, go for an earthy, passionate, vivid color. Terra-cotta seemed just right. We ordered samples from the stucco supplier, and I finally settled on what the plaster people considered the limit of pigment per batch that would not jeopardize the chemistry. We applied test sections on the house. Linda Klein and I drove up from Pasadena to approve the decision. It was risky, but we loved the color, and told the plasterers to go for it.

Completed, the house looked absolutely awful. The sheer intensity of those huge stuccoed walls in the sun was blinding. The white house next door had turned pink in their reflection. What little wood detail there was on the exterior—rafter tails and windows—was swallowed up by this pulsing terra-cotta glow. Stewart and the crew were totally depressed.

My clients agreed to split the cost of redoing the exterior coat. We thought about painting over the offending surfaces. But the likelihood of having patches of paint blasted away by wind-driven sand was reason enough to have the finish coat of stucco scraped off and a fresh one applied. The plasterer was upset with the results, too, and gave us a good price on redoing the finish coat. We recruited a small army of laborers and armed each with a floor scraper and dust mask. After several days' work we had removed enough stucco to bond another finish coat, although the adhesion would not be as good as the first time. The next coat matched the color of the sand.

Interior finish—The interior walls and ceilings were sheathed entirely with ⅝-in. drywall. I would have preferred the same 1x6 T&G on the ceilings that we used under the eaves, but the cost would have been too high. To blend these sheet materials together and evoke Craftsman period interiors, we used the skim-coat technique we'd learned in restoration work.

First, we taped all the joints, and gave the rock one coat of mud in conventional fashion. Then we mixed latex drywall joint compound with #60 silica sand and troweled the stuff over the walls and ceilings. Working an area of about 3 sq. ft. at a time, we parged the mud/sand mixture onto the rock to a consistent thickness of about ⅛ in. When the mixture was about to set up (it starts getting a little leathery at this point), we troweled the surface with a moistened sponge float to bring the sand to the surface and give the walls an even texture. Only one coat was needed, and the result looked just like three-coat plaster. The ceilings and walls in the kitchen and baths were finished smooth (without the sand) because a rough texture is inappropriate for these rooms.

Casings, baseboards and continuous header banding were to be made of the least expensive paint-grade wood possible. The pine that Stewart bought arrived on the job and looked much better than we expected, so we resisted the plan to paint it white. The plaster walls and ceilings were going to be medium grey. We needed a light, off-white trim color that could be as easily applied as an oil finish. But oil alone gives pine a wretched yellow color, so we bought some titanium white tint from the paint store, mixed it with Waterlox, a polymerizing tung oil (Waterlox Chemical and Coatings Corp., 9808 Meech Ave., Cleveland, Ohio 44105), and wiped it on a sample board. The white killed the yellow, but let the grain and character of the wood remain. This technique also reduced labor. In one application we could finish the pieces of precut trim before installing them.

We left the header strips ¾ in. thick, milled the casings ⅝ in. thick and the baseboards ½ in. thick. There is a slight step from one trim surface to another at each connection. Stewart considered various options for joining the trim boards. He worked out a quick and visually pleasing way to connect pieces with simple butt joints, which depends on adjoining members being of different thicknesses. Where, say, a header strip meets a window casing, the underside of the thicker casing is rabbeted so the top end overlaps the casing by ½ in. The corners of the header are slightly rounded and all the edges are radiused with a ¼-in. piloted router bit. Other connections, like side casing to head casing, are plain butt joints with the edges of the thicker member radiused. Trim pieces that join at outside corners are handled by letting one piece run ½ in. beyond the other and rounding over the proud member's corners and edges.

This simple method of joining trim saved a lot of time because the carpenters didn't have to fiddle with cutting and fitting mitered joints. And true to the Craftsman look, there's an interesting layering of planes where pieces of trim come together. It's something you just don't get with mitered moldings.

Continuous header strips above doors and windows were another common device in Craftsman homes. They were used to unify the space, to tie things together visually and to scale down rooms to make them seem more intimate. In many Craftsman homes the paint color would change at this juncture. The upper color would wrap over the ceiling down onto the wall, where it would stop at the header strip. We used the same grey-green hue above the header strips, but lightened it to brighten the ceiling. With the 5½-in. white band separating the two, people seldom notice the the change in tint and tone, yet the desired effect is achieved.

The fireplace is a prefabricated unit with a triple-wall metal flue. To retain as much view as possible from the kitchen and dining room, the flue angles back to the wall once it comes out of the firebox. Stewart worked out a series of stepped gypboard shelves to box this pipe in. Since the code prevented our using wood this close to a flue, Stewart couldn't build a 2x framework to hold the rock. So he fabricated the entire stepped flue box by joining the pieces of gypboard with 90° outside corner bead. He held the bead on the inside of the corner and attached the rock with drywall screws, driving them through the rock and into the metal bead. Wood shelves and trim were added on top. Tile facing around the firebox and a raised hearth extend the line of the baseboard and increase the fireplace's appearance of mass—a quality that seems so comforting about masonry.

As completed, the house is bright and feels ample despite its small size. Interlocking the spaces was definitely the right decision. The realtors complained about the windows being too fussy and have asked how much it would cost to put in a nice aluminum sliding door to face the beach. We like the windows and the unexpected glimpses of the ocean one gets as you move through the building.

Ventilation works well with the transoms above the west windows and an operable skylight at the top of the stairwell. As the house heats up in the afternoon, the transoms can be opened to the prevailing westerly breezes, and the skylight exhausts the warm air from the highest point in the house.

Often when I show people the house, they aren't sure whether it's new or an old one in good repair. It fits easily into the neighborhood, and easily into one's notion of what a beach house should be. □

Tim Andersen, of Pasadena, Calif., runs an architectural practice, teaches design history at Art Center College and is chairman of the Pasadena Cultural Heritage Commission.

Living-room design. Wood was carefully selected for figure and color. In the living-room, darker wood flanking the entry accentuates the swan-neck joints of the casing and contrasts with the horizontal trim (detail photos, far right and facing page). Inherent characteristics of the wood often suggested the designs, as in the trim leading up to the skylight (above). This trim balances the arched doorway, not centered on the wall, and helps to reduce the effect of a sharply pitched ceiling. All edges were rounded over with a router. Doors to the living room hang from concealed tracks mounted on the hallway side of the opening. Door handles (right) are carved ebony. Above the mantel, four bookmatched panels of padauk form a backdrop for a 14th-century Tibetan sculpture. The panels, planed to ¼-in. thickness, were laminated to a ½-in. Baltic birch plywood substrate for stability.

Craftsman Remodel

A Greene and Greene inspired living room and bedroom

by Peter Malakoff

The Rosestone remodel was a project with a modest beginning. Doug and Beth Rosestone originally asked me to make a built-in stereo cabinet, and we arranged a meeting to discuss the design. After a short tour of the house, we sat in the living room where the proposed cabinet would go.

As I looked around, I began to realize that the room wasn't really suited to showcase the cabinet the clients had in mind. The greatest offender was the fireplace, which dominated the room with a synthetic lava-rock facade. The room was a typically large, contemporary living area in California's diluted "Spanish" style. Dark-stained, rough 4x10 fir beams ran from ridge to eave supporting a fir 1x6 T&G ceiling.

After some hesitation, I voiced my opinion about the room's unfortunate design. This was the beginning of a remodeling project that eventually transformed the Rosestones' entire house. The living room and bedroom alone took seven months to complete.

Designing the living room—After our first meeting, Doug Rosestone asked to see a drawing for remodeling the living room. We agreed on three design goals: to create an arched entry to the living room, to remove the lava-rock veneer from the fireplace area, and to fashion a display for a 14th-century Tibetan tantric sculpture. To help us visualize these changes, we chose Eli Sutton as designer.

Some of the ideas and motifs we used came from Randell Makinson's *Greene & Greene: Architecture as Fine Art* (Peregrine Smith, Box 667, 1877 E. Gentile St., Layton, Utah 84041). We turned also to examples of *Sukiya* (teahouse) architecture for inspiration. Both styles are marked by elegant simplicity, a respectful attention to detail and an appreciation for architecture's ability to enhance the quality of life. Sutton and I started roughing out our ideas on paper. We sketched things out in a rush of creativity, and completed the basic design in a matter of minutes. The sketch we drew in that one session became, with surprisingly few alterations, the finished design for the room.

It is extremely difficult to bid large-scale projects when they involve a lot of custom work, so we proposed to work on a time-and-materials basis. Since the Rosestones had by then left on vacation, we mailed them elevation drawings of the project, along with photocopied illustrations from the Greene and Greene book, to give them a feel for what we had in mind. Fortunately, we didn't have to wait long for a reply. Rosestone phoned from Hawaii, asking us to start immediately on the living room, and to develop a design for the bedroom in the same style.

Making an on-site shop—In a project where 20-ft. boards and multiple milling operations are common, a spacious on-site shop is essential. So the first thing that we did was to convert the Rosestone garage into a temporary workshop. This gave us room to install several power tools—a jointer, a thickness planer, a table saw, a bandsaw, a radial-arm saw and a drill press, which doubled occasionally as a lathe. One of the hardest working and most used tools of the project was a Hitachi B600A bandsaw. We also built wood-storage racks, two workbenches and an outfeed extension table for the table saw. Still, with up to ten carpenters at work and several activities in progress at one time, there were days that the garage shop seemed small.

Building the archways—The arch that forms the entry into the living room was to be something special, with swan-neck joints on the side casings. The side jambs and head jamb are one continuous lamination (photos facing page). I saw the arch as a sign of our craftsmanship, as it would frame the view of the work that was taking place in the living room. Cabinetmaker Mark Berry and I had both done bent laminations before, but we'd never tackled a project of this size. Though we installed several arches in the house, the first one proved to be the teacher.

The arch for the jambs was made of padauk, purchased in standard 4/4, 6/4 and 8/4 stock and resawn to a rough thickness of 3⁄16 in. After every pass through the bandsaw, the stock was run through a thickness planer to smooth the sawn surface. This yielded a true face and eliminated irregularities that would compound themselves as successive cuts were made. A second pass through the planer brought the resawn boards to their final ⅛-in. thickness.

Because padauk is an oily wood, we had to clean the mating surfaces with acetone to ensure proper adhesion. The marine-grade plastic-resin glue we used has a long working time and makes a strong bond. The laminae (plies) were 6 in. wide and 15 ft. long, and since the arch opening was approximately 18 ft. around, we alternated 15-ft. and 3-ft. lengths from side to side on adjacent plies. For the last ply—the visible, inside surface of the arch—we used a single 18-ft. long piece. Though at first it had seemed

Photos: Fritz Hoddick

A number of doorways in the house received laminated arches for jambs, which were built on site. Waxed paper kept the padauk laminations from sticking to the forms. Clamping began at the apex of the form, and proceeded gradually down both sides.

logical to build the laminated arches in place, using the wall framing as our pattern, we found the technique troublesome, and soon concluded that it would be better to prefabricate them.

On subsequent arches we built a form of ¾-in. plywood and 2x4s to the exact specifications of the opening, as shown in the photo above. Because of the time involved in spreading the glue and clamping up, five layers were the most we could glue up at once. Additionally, the stress of bending more than five layers over the frame was too much for our clamps to overcome. When the tenth ply was done, we removed the arch from the form and cleaned up its edges with a power planer. It took us about five days to build the frame, laminate a typical arch, install and finish it.

Work in progress—When the Rosestones returned from their vacation, their home was in the "partial" stage—partially demolished, partially completed and partially livable. In the living room it was difficult to distinguish between what we were building and what we were tearing out. Their garage had become a workshop. Dust from drywall and wood covered the living quarters, and most of the electrical outlets and fixtures were disconnected.

It took the Rosestones some time to comprehend what was taking place. Soon however, they began to ask questions, make suggestions and even compliment the crew. At this stage, the creative conversation between the craftsmen, designer and clients truly began to work for the project.

Selecting wood by color and grain—One part of the project I thoroughly enjoyed was selecting the woods. We found that mahogany was best for the Rosestone remodel. It came in long, wide boards and was easy to work. I spent hours sorting through piles of Honduras mahogany at a local yard I use frequently. The selection process wasn't always clear cut, and my recommendations weren't necessarily considered final by the other craftsmen. We often argued the merits of one piece over another.

Each board in the living room was selected for its color, figure and texture. For example, there's a piece of trim near the skylights that takes a sharp turn. We found a board with a rare right-angle bend in the grain to use there. The trim over the mantel has an attractive elliptical figure to the grain.

Since I didn't have unlimited funds for the project, the cost of wood always concerned me. The project required many pieces that were 15 ft. to 20 ft. long and 10 in. to 14 in. wide, and I was always looking for ways to maximize quality and minimize the cost. In searching for ebony, which we used for some of the trim and detailing, I found that it was best to buy whole logs from Jordan International (1303 Elmer St., Belmont, Calif. 94002) and mill the lumber to our own specifications. In spite of a 60% waste factor by weight, we were able to get better-quality lumber at a lower cost than if we had purchased already dimensioned boards.

When we chose wood there was a strong temptation to think, "So what? How much difference can it make?" But now that all those laboriously selected boards are together in one composition, they have a synergistic effect—the sum is more than the simple aggregate of individual elements.

Working with illusions—The living room's existing 4x10 fir beams posed a problem. Their rough texture, sharp edges and dark brown stain clashed with the mood we were creating. On the other hand, the exposed beams fit the living room's new style. The existing beams simply had to be upgraded. We did this by casing them in mahogany on three sides to give the illusion of solid mahogany beams. The U-shaped casings were prefabricated in the shop like all the other trim elements. Because they had to fit perfectly against the walls, we cut the ends with a *ryoba* saw (a Japanese double-edged handsaw with ripping teeth on one edge and crosscutting teeth on the other edge).

The steep pitch of the ceiling in the living

A new bedroom ceiling. **The new ceiling was constructed with interlocking joinery (above), with a minimum of mechanical fasteners and glue. The gable-end screens were made from interlocking fir lattice, shaped on a router table to resemble bamboo, and tied with alternating red and black threads (facing page).**

The aluminum mullions of the bedroom skylight were sheathed in wood, left, to match the interior trim. The *shoji* panels for the new ceiling were installed from above by loading them through the skylight opening.

From *Fine Homebuilding* magazine (October 1985) 29:56-61

Shoji operating system

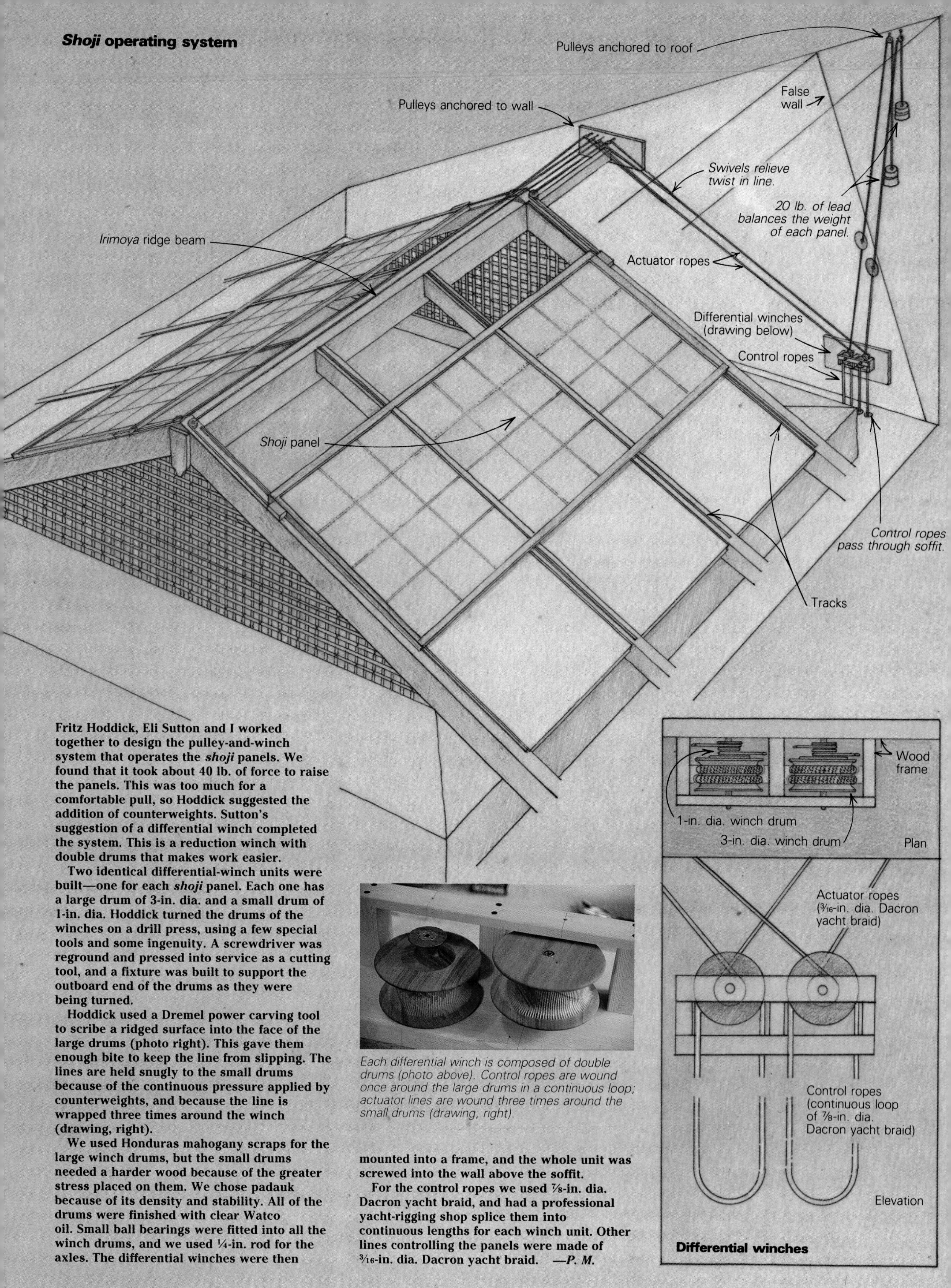

Each differential winch is composed of double drums (photo above). Control ropes are wound once around the large drums in a continuous loop; actuator lines are wound three times around the small drums (drawing, right).

Fritz Hoddick, Eli Sutton and I worked together to design the pulley-and-winch system that operates the *shoji* panels. We found that it took about 40 lb. of force to raise the panels. This was too much for a comfortable pull, so Hoddick suggested the addition of counterweights. Sutton's suggestion of a differential winch completed the system. This is a reduction winch with double drums that makes work easier.

Two identical differential-winch units were built—one for each *shoji* panel. Each one has a large drum of 3-in. dia. and a small drum of 1-in. dia. Hoddick turned the drums of the winches on a drill press, using a few special tools and some ingenuity. A screwdriver was reground and pressed into service as a cutting tool, and a fixture was built to support the outboard end of the drums as they were being turned.

Hoddick used a Dremel power carving tool to scribe a ridged surface into the face of the large drums (photo right). This gave them enough bite to keep the line from slipping. The lines are held snugly to the small drums because of the continuous pressure applied by counterweights, and because the line is wrapped three times around the winch (drawing, right).

We used Honduras mahogany scraps for the large winch drums, but the small drums needed a harder wood because of the greater stress placed on them. We chose padauk because of its density and stability. All of the drums were finished with clear Watco oil. Small ball bearings were fitted into all the winch drums, and we used ¼-in. rod for the axles. The differential winches were then mounted into a frame, and the whole unit was screwed into the wall above the soffit.

For the control ropes we used ⅞-in. dia. Dacron yacht braid, and had a professional yacht-rigging shop splice them into continuous lengths for each winch unit. Other lines controlling the panels were made of 3/16-in. dia. Dacron yacht braid. —*P. M.*

Drawings: Victor Lazzaro

room was something that we wanted to minimize. The accent trim and soft edges of the mahogany beam casings helped to do this, as did the skylight and *shoji* screens. The living-room skylights were made simply by inserting a ¼-in. Plexiglas panel under the shingling, and caulking around the outside edges (photo, p. 152, bottom left). This solution proved to be simple, economical and effective. The *shoji* soften natural light as it enters the room and blunt the sharp peak of the ceiling. Between the skylight and the *shoji* are concealed lights for illuminating the room at night.

The fireplace, mantel and hearth—After the artifical lava stone was stripped away from the fireplace, the work of creating a focal point for the room began in earnest. Rosestone had asked us to create a place for a work of 14th-century Tibetan sculpture. I was aware of the significance of the piece and wanted to find the right symbolic setting for it. While searching for wood I came across a stunning piece of padauk. Its fiery grain pattern immediately suggested a design. I could see that by bookmatching the board we could create a dramatic backdrop for the statue (top photo, p. 150).

In bookmatching, it often happens that the pieces lose their precise symmetry because the contiguous layers of grain become further removed from each other through resawing and sanding. In this case, the imperfections led to another unifying element for the design of the project. A flaw appeared in a conspicuous spot at the top of the two center panels. To mask this imperfection we decided to inlay an ebony circle over the joint in the panel. The resulting patch appears directly over the sculpture and accentuates the drama of its setting.

A change of plans in the bedroom—The bedroom originally had a flat 8-ft. ceiling, and our first design called for tearing it out. A new ceiling incorporating delicate *shoji* would filter light into the room from new skylights above. After eight days of work, we had torn out the old ceiling and almost completed the new one. But Rosestone decided that he wanted more than just a diffused lighting effect from the *shoji* panels; he wanted to see the starscape at night. He was adamant on the point, so we began to design a movable *shoji* panel.

Everyone was enthusiastic about movable *shoji*, but the mechanical details gave us problems, and solving them was going to cost a lot of money. Even so, Rosestone insisted on seeing a new design and an estimate of the cost. He liked it, and we had to tear out the partially completed ceiling and start over from scratch.

Sutton's design created a false ceiling similar in shape to an *irimoya*, which is a Japanese roof like a gable roof. When the *shoji* panels are opened, they slide down into pockets between the false ceiling and the roof (photo, p. 153). The soffit around the room appears to carry the weight of the ceiling and roof, but it doesn't. The load is actually carried by a ledger screwed into the wall studs.

Building the bedroom—The new ceiling was made entirely in our shop after a day spent taking and double-checking all the measurements in the bedroom. The most critical part of the project was keeping the angle of the A-frame perfectly aligned with the pitch of the roof. Once constructed, we dismantled the ceiling and reconstructed it in the bedroom. This was relatively easy since the structure depended almost entirely on interlocking joinery for its strength, not on glue or hardware.

Bird's mouths on the rafters rest on ledgers lag-bolted to each of the walls (detail photo, p. 152). The ceiling panels are ¼-in. drywall laminated to 1-in. Styrofoam insulation. They simply drop into the ceiling grid, with the pre-textured and painted surface of the drywall facing down. A double-insulated glass skylight floods the room with light. Its three aluminum mullions were covered with wood because we didn't want any metal showing.

Near the ridge of the roof, we installed a heat-exhaust vent, which was made from a single piece of kiln-dried fir. To build it, we cut kerfs in one direction with a radial-arm saw, then cut evenly spaced dadoes on the other side. The intersection of these two cuts created rectangular openings in the wood. Finally, using a small Japanese roundover plane, we eased the longitudinal edges of the gridwork. An exhaust fan in the roof draws hot air from the ceiling.

There are 28 15-watt bulbs mounted in the soffit around the room in a parallel circuit. When the dimmer is turned up, each bulb casts a scallop of light in the segmented bays of the soffit (photo below). It is as if the room were candlelit, and the mood changes dramatically. Stereo speakers are mounted out of sight in the soffit. Without any elaborate planning, it happened that the room has excellent acoustics.

Pulleys and rope—The greatest mechanical challenge of the project was designing a system to open and close the *shoji* panels to reveal the skylight. This task fell to Fritz Hoddick, who was looking for a job as a computer programmer when we hired him on as a temporary carpenter. His technical background proved invaluable in working out and constructing the system of ropes, pulleys and counterweights for the movable *shoji* panels (see the facing page).

Baseboard and door—The baseboard and door trim in the bedroom have an integrated design. Two exposed ½-in. grooves cut into the baseboard run the perimeter of the room and continue into the door trim to tie all the trim work together. To construct the door we built a framework of kiln-dried fir, and then laminated ¼-in. thick fir planks horizontally onto that base. Each of the planks is slightly different in width, which gives the door a subtle individuality.

We are very pleased with the way this project turned out and with our collective experience on this job, which formed the basis of a fruitful association. But the Rosestone remodel is not quite finished yet. That stereo cabinet has yet to be built. □

Peter Malakoff, formerly a principal in Hoddick Berry & Malakoff, is now an independent woodworker in San Rafael, Calif.

Twenty-eight 15-watt bulbs are concealed in the soffit and controlled by a single dimmer switch. The effects are particularly dramatic when the *shoji* panels are closed.

Gypsy Wagon

Truck parts, a load of cedar and salvaged furniture make a home on wheels

by Daniel Wing

My wife Dina Dubois and I began talking about building a Gypsy wagon when we first met, but it seemed an improbable dream. Our early sketches were based only on childhood memories of Mr. Toad's Gypsy caravan in *The Wind in the Willows*, and on my experience in building utility trailers and rebuilding old cars and trucks. Then we met Jim Tolpin of Port Townsend, Washington, who is probably America's only professional builder of *vardos*, which is what Gypsies call these wagons. He showed us a wagon he had built and pictures of several others. We decided to build one ourselves, with occasional phone calls to Tolpin for advice.

In the summer of 1987, we hitched up the wagon to our '52 Ford pickup and started a journey from Seattle to our home state of Vermont, by way of New Mexico (photo right). For a month and a half, we pulled the wagon across 4,500 miles of deserts, mountain ranges and mall parking lots. We were surprised at how easily the wagon towed and how stable it was. We were also surprised at how cozy it was at night with the woodstove cooking and the moonlight coming through the roof, and how nice it felt to eat at "home" instead of in a fast-food joint.

Photo: Dina Dubois

Built in the traditional ledge style, this modern-day Gypsy wagon takes a cross-country journey hitched to a rebuilt '52 Ford pickup.

But we were not prepared for the biggest surprise, which was the deep interest and instant affection the wagon triggered in almost everyone who saw it. Hundreds of people went out of their way to talk to us, and some invited us to park the wagon long enough to get acquainted.

Traditional Gypsy wagons—The Gypsy wagon flourished in England from the construction of modern roads in the 1850s until the widespread use of motor transport in the mid-1920s. English wagons were not built by the Gypsies, or travelers, as they called themselves, but by specialized wagon-makers. These wagons were constructed as small timber-frame houses, about 10½ ft. long and 7½ ft. wide. The wood for frame and skin was fir or pine, with curved parts made from ash. Joints were pegged mortise-and-tenon and walls were finished with narrow shiplapped "pennyboarding." Summer beams supported floorboards in the wagon proper, as well as the footboard for the driver. Roof ribs were sawn into curves, decked with wood and covered with paint-impregnated canvas. Some wagons incorporated a clerestory, or *mollicroft*, which provided light and ventilation. The mollicroft allowed shorter roof ribs but required extra joinery.

Mahogany cabinets were built into the sides of the wagon, and two beds into the back. The woodstove was on the right side of English wagons, where its chimney was less likely to snag branches. Side and rear double-hung windows could be covered from the outside by sliding shutters held in place by trim boards running the length of the wagon. Ornate embellishments were often applied over the outer surface of the wagon between the ribs, and exterior structural members were decoratively chamfered. No expense was spared in painting and gilding the finished wagon.

The ledge wagon—The use of ledges was one of the traditional methods of making a deep and wide wagon. A ledge is a lengthwise board that creates an inset to provide clearance for the wheels. On the inside of the vardo, the top of the ledge is 18 in. above the floor, and forms a bench wherever it is not covered up by cabinets, bed or stove.

When we were ready to plan our own wagon, Jim Tolpin gave us the address of John Thompson, an English modelmaker who sells detailed plans of a wide range of reproduction horse-drawn wagons and farm equipment (1 Fieldway, Fleet, Hampshire, England). From these plans we selected a Dunton and Sons ledge wagon, first built in Reading, England, in 1914. We built a vastly simplified version.

We wanted the height of the wagon to be less than 10½ ft. The ledge design permits comfortable headroom and leaves enough space below the ledge for 1932 Ford wheels and cycle-style fenders, while providing side clearance for the front steering motion of the wheels.

Building the undercarriage—To make the most room for our transverse double bed, we figured on an 8-ft. wide wagon, the maximum width for highway travel. I used front axles from 1959 Ford pickup trucks for both front and rear axles of the wagon. Before removing the superfluous tie rod on the rear axle, I welded the axle and spindles together in the straight-ahead position and widened the axles with a steel box channel. For flexibility on the road, I made both a hay-wagon hitch and a stiff hitch, and installed quick disconnects in the trailer light and brake connections in anticipation of making hitch changeovers.

The front wheels of a hay-wagon hitch are free to steer on the axle, but are connected to each other with a tie rod. One wheel is also connected by a drag link to the shorter section of a two-part tow bar. The tow bar can swivel side to side and pivot up and down on pin connections. When the horse (or tow rig) turns left, the drag link forces the wheels to turn left, and the wagon follows smoothly. An experienced driver can back up this type of wagon, and the hitch is stable and easy to

From *Fine Homebuilding* magazine (December 1988) 50:87-89

maneuver at low speeds, even on ruts and rocks. At speeds over 25 mph, the wagon would have a tendency to wander from side to side, although this motion would not be felt in the tow rig. Careful adjustment of steering geometry and alignment might eliminate this problem, but I simply avoid this hitch for highway use.

The stiff hitch uses only the long section of the tow bar, which must be rigidly attached and triangulated to the front axle of the wagon. The tow bar can pivot up and down, but not from side to side. The wheels remain connected to each other with only a tie rod. Some extra caster is set into the alignment of the front wheels to help them follow passively behind the tow rig, and I put a hydraulic steering damper on the tie rod to prevent shimmy. This hitch can't be backed up more than a few feet, even if the trailer and tow rig are in a straight line. It's also relatively unstable at low speeds because one rock can knock the wheels out of the line of travel. But at speeds over 15 mph, the stiff hitch is stable and impervious to cross winds and road irregularities.

We used a hydraulic brake system with a surge mechanism, in which a master cylinder on the trailer is activated by the force of the trailer surging against the hitch when the tow rig is decelerating. To avoid having to work on the brakes in the future, I filled the system with silicone brake fluid, also called DOT 5, which is an expensive but permanent means of preventing corrosion in the hydraulic cylinders.

I used pickup-truck springs on the front axle and lightweight trailer springs in the back and added gas-filled shock absorbers on the front to control any tendency of the wagon to porpoise.

A wood-and-metal subframe—Rather than support the wagon on a separate metal subframe, which is standard practice for utility trailers, I first built a wood frame of 2x6 and 2x4 cedar. Then, using the cedar frame as a guide, I made a flat metal frame of welded bar stock and bolted it to the bottom of the cedar frame. On top of the cedar frame, I built the wagon floor from salvaged ⅜-in. T&G oak, which I stripped and gave a Swedish finish before building the wagon walls.

I built the lower wagon walls in ledge style. To make the sides, I glued and nailed a 2-ft. by 14-ft. length of marine plywood to each long outside face of the cedar frame. To make each ledge I edge-glued two 2x6s and a 2x4 together, then used a router to cut a slot on the underside to house the top edge of the marine plywood (photo below). My next step was to attach the subframe and lower-wall assembly to the undercarriage. I welded and bolted spring perches to the wood-and-metal subframe and bolted the perches to the springs. I was ready to frame the upper walls.

Woodwork and structure in one—Traditional wagons were held together with joinery, not glue, and they were known to loosen up considerably under the stress of negotiating 19th-century roads. Gypsies traded and sold their wagons among themselves frequently, and it was traditional to burn a vardo when its owner died. A wagon's durability was considered secondary to its ornamentation. For us, however, the prospect of structural failure at highway speeds was daunting. Although Western red cedar is an enjoyable wood to work and live with, it's not terribly strong. I elected to glue and mechanically fasten most joints rather than rely entirely on joinery. I chose stainless, silicon bronze, brass, copper and hot-dipped galvanized fasteners, all of which are easy to find in Seattle, one of the wooden-boat building centers of the world.

Using cedar as the basic wood for the wagon was an easy choice—in Seattle, it's no more expensive than an equivalent grade of any other wood. And, it was available in a variety of thin, narrow dimensions. Early in construction I was offered 1,700 lineal feet of $^{11}/_{32}$-in. by 2⅛-in. clear T&G cedar boards—planed on one side and matched, and in lengths of up to 16 ft.—at 20¢ per linear foot, provided I took it all. With this stock I was able to side and roof the entire wagon. I used Douglas fir, because of its superior strength, for roof ribs, plates and mid-rails below the windows. I sealed the cedar and fir with Deks Olje no. 1, a penetrating marine varnish that can be reapplied at any time (The Flood Co., Marine Products Division, P. O. Box 399, Hudson, Ohio 44236).

We had originally designed an intricate Art Nouveau door for the wagon, but a $3,000 estimate for the work from a cabinet shop forced a design change. I made a Dutch door from an Old English row-house door by sawing it in half, then cut a curve in the top with a saber saw.

To make a curved piece that would act as both door lintel and front wall plate, I laminated a plywood core between 1x cedar stock, then cut a compound curve in the bottom to match the door and a simple curve along the top (photo below). I made the jambs from 2x4s and 4x4s, cut an open mortise into the bottom of each 4x4, slipped the post over the front of the wagon frame and bolted it in place.

Framing for the back wall included 4x4 corner posts, a 2x4 window header, 1x1½-in. studs and a sheet of plywood across the whole lower end to tie the two sides together and strengthen the whole wagon. I cut a hole through this sheet for access to storage, but left enough intact for stiffness. I fastened the sidewall studs to the plates and to the ledge with one 3½-in. bronze square-drive screw at each end. Single-shouldered tenons were cut into the studs above and below the window, and the studs were then screwed to the window frame. The 2x4 side wall plates are stiffened by the fir trim board that holds the top of the shutters. I used epoxy for the wood joints that needed the most stiffness (such as the joint between the ledge top and the plywood web), but for other joints, I used either premium-grade construction adhesive or marine sealant for ease of application and flexibility. The joints are strong and water-resistant.

Light inside and out—Neither time nor skill permitted me the chance to build a mollicroft, so I glue-laminated two 1½-in. wide strips of ½-in. Douglas fir over a curved form and screwed the ends of each curved rib to the wall plates. To increase stiffness, I added one especially deep laminated roof rib in the

A ledge built of horizontal edge-glued cedar 2xs and sides formed by marine plywood make up each lower wall, which is glued and screwed to the subframe. Upper wall framing is 1x2 and 4x4 cedar, but the roof ribs are made of Douglas fir for strength.

center of the vaulted roof. Had I the chance to build another vardo, I'd stiffen the ribs a bit. I'd also use a few more deep ribs to strengthen the roof.

I had decked half the roof with cedar before Dina realized that the interior of the vardo was going to be too dark. Figuring that a skylight of some sort would solve the problem, I left undecked a large area in the middle of the roof. We had a large sheet of polyester-reinforced vinyl awning material made into a roof cover, as on a traditional canvas-roofed wagon. This material is available in many colors and in clear, which we chose because it looks like white canvas from 2 ft. away. We threaded gold-colored nylon rope through grommets in the material and through screw eyes attached to the wagon eaves (photo right).

Every state requires clearance marker lights and tail lights, as well as safety chains, fenders (if the wheels are outside the body of the trailer) and brakes (if the trailer is above a certain weight). Conventional lights would look out of place on a vardo so I recessed the bases of the marker lights, letting only the lenses show. A friend who works in a boat yard gave me two old brass salon lamps, which I rewired and fitted with colored plastic lenses—amber in the sides and front, and red behind (photo, p. 156). To make the red lenses brighter, I cut out a portion in the center of each lens and glued a conventional prismatic trailer lens in place.

The partially decked roof allows light in through the reinforced-vinyl roofing. A built-in bed contains storage space and sliding table tops (photo at top). Outside, steps are removable, and the landing swings up for travel (photo above).

Fitting up the wagon—Dina found two small cupboards to fit on the ledge beneath a large antique European cupboard. For the table leaves that slide under the bed, she found sycamore boards that had been broken off of an Old English dresser (top photo). I cut up salvaged six-panel fir doors to make doors for cabinets in the kitchen and below the bed. Old brass refrigerator latches hold the kitchen cabinet doors in place.

I had already used the headboard and footboard of an old Danish twin bed for window frames in the side walls. Making wooden sash would have been difficult and would have taken up too much window area, so I used copper came to make the side windows and soldered the edge of the copper sash to brass piano hinges to allow the windows to open inward.

Making the windows gave me enough confidence to fabricate copper countertops and to solder-in a brass sink (top photo). A 95-lb. cast-iron wood-burning cookstove stands away from the wall and kitchen cabinets, which are shielded by sheet copper. Even the walls next to the stovepipe stay cool, but we are going to make a metal shield for the stovepipe anyway.

Under the floor is a 24-gal. water tank. Outside, in one of the storage boxes tucked between the wheels, is an RV/marine battery that trickle-charges as we drive. The propane tank is in its own compartment under the bed, vented to the outside through large holes drilled in the floor (propane gas is heavier than air).

Second thoughts—It took five months of full-time work to build the wagon, and about $7,500. I'll probably never build another one, but if I did, I'd do some things differently, such as add soft shock absorbers to the undercarriage and widen the ledge to 18 in. for a more comfortable seat. But Dina and I love the wagon as much as everyone else seems to. Now all we have to do is learn to drive horses and tell fortunes. □

Daniel Wing is a physician and tinkerer in Corinth, Vermont.

Index